ADEX Theory

How the ADE Coxeter Graphs Unify
Mathematics and Physics

K&E Series on Knots and Everything — Vol. 57

ADEX Theory

How the ADE Coxeter Graphs Unify Mathematics and Physics

Saul-Paul Sirag

NEW JERSEY • LONDON • SINGAPORE • BEIJING • SHANGHAI • HONG KONG • TAIPEI • CHENNAI • TOKYO

Published by

World Scientific Publishing Co. Pte. Ltd.
5 Toh Tuck Link, Singapore 596224
USA office: 27 Warren Street, Suite 401-402, Hackensack, NJ 07601
UK office: 57 Shelton Street, Covent Garden, London WC2H 9HE

Library of Congress Cataloging-in-Publication Data
Sirag, Saul Paul.
ADEX theory : how the ADE coxeter graphs unify mathematics and physics / by Saul-Paul Sirag.
pages cm. -- (Series on knots and everything ; vol. 57)
Includes bibliographical references and index.
ISBN 978-9814656498 (hardcover : alk. paper)
1. Coxeter groups. 2. Coxeter graphs. 3. Knot theory. 4. Mathematical physics. I. Title.
QA177.S57 2016
512'.2--dc23

2015026760

British Library Cataloguing-in-Publication Data
A catalogue record for this book is available from the British Library.

Typeset by Stallion Press
Email: enquiries@stallionpress.com

Printed in Singapore

To Mary-Minn

Preface

This book is meant to be read by those who enjoy the interplay of mathematics and physics. Going back to the age of Archimedes, all of the advances in physics have been accomplished by a deep relationship between these two seminal fields of human knowledge. Indeed, many of the advances in mathematics have been inspired by attempting to solve problems in physics. Perhaps Newton is the prime exemplar of this approach. His attempt to understand gravity led to his development of the calculus, differential equations, and other areas of mathematics.

Physicists today are engaged in the very difficult, but enthralling, effort to unify all the forces. By far the most promising path seems to be string theory, which has evolved into a vast enterprise of mathematical discovery and application to the quantum theory of gravity unified with the Standard Model gauge forces.

For several decades now, I have been convinced that the organizing principle for this physical unification is the vast unification process afforded by the unification of mathematical structures inherent to the ADE classification of twenty-some very different mathematical objects. I have been impressed by the fact that all of these mathematical structures seem to have applications in physics — especially the physics of string theory.

Accordingly I have used the term ADEX theory to refer to the study and application of all the ADE-classified mathematical objects. The X should stand for the structure underlying all the ADE classified objects. Each of these mathematical objects is a separate window into the underlying structure. It seems that this underlying structure must be the quantum physical world as elaborated in string theory.

In developing my approach to string theory by way of ADEX theory, I have been encouraged, inspired, and instructed by many colleagues (including those who argued against my views), I would like to acknowledge especially.

Nick Herbert, Jeffrey Mishlove, Elizabeth Rauscher, Jack Sarfatti, Creon Levit, Tony Smith, and most of all, Louis Kauffman, who has over the years been prompting me to write this book.

Note to the Reader

For references in the text indicating papers and books to be found in the Bibliography, I have used the convention of square brackets with the authors name and date of publication, and occasionally a specific page number.

For example: [Yau, 1977], [Born, 1971, p. 158].

Some of these references occur also in the Glossary, whose terms I have described rather fully, rather than more briefly.

Contents

ADEX Theory

How the ADE Coxeter Graphs Unify
Mathematics and Physics

Chapter 1

Introduction

In 1974, Arthur Young, at the Institute for the Study of Consciousness in Berkeley, California, asked me as his research associate to work out the symmetry group of a toy tetrahedron. This turned out to be $\mathcal{S}_4$, the symmetric-4 group, which is the set of all 24 permutations of four objects (for example, the four vertices of the tetrahedron). The rotations of the tetrahedron corresponded to the 12-element normal subgroup of $\mathcal{S}_4$, called $\mathcal{A}_4$, the alternating-4 group, or the tetrahedral group. The $\mathcal{S}_4$ group also includes reflections of the tetrahedron and is called the octahedral group because it corresponds to all the rotations of the octahedron (or the cube, which is its dual).

These group structures entail the basic themes of this book: symmetry, rotation, reflection, permutation, subgroup, duality, commutativity and non-commutativity, cosets, mappings, representations, and character tables. Underlying it all is the interplay between geometry and algebra.

Moreover, the tetrahedral ($\mathcal{T}$) and octahedral ($\mathcal{O}$) groups correspond to two of the ADE Coxeter graphs, E_6 and E_7, via the McKay correspondence groups $\mathcal{TD}$ and $\mathcal{OD}$, which are double covers of $\mathcal{T}$ and $\mathcal{O}$.

The ostensible connection between group structures and consciousness (a very controversial topic) was via Arthur Eddington's fascination with symmetry groups as fundamental to physics. In his book, *The Philosophy of Physical Science*, Eddington [1939] wrote:

"The recognition that physical knowledge is structural knowledge abolishes all dualism of consciousness and matter. Dualism depends on the belief that we find in the external world something of a nature incommensurable with what we find in consciousness; but all that physical science reveals to us in the external world is group-structure, and group-structure is also to be found in consciousness. When we take a structure of sensations in a particular consciousness and describe it in physical terms as part of the structure of an external world, it is still a structure of sensations. It would be entirely pointless to invent something else for it to be a structure of."

Since 1939 the importance of group theory in physics has grown by leaps and bounds, and is now at the very center of advances in unified field theory and other areas of theoretical physics. Indeed, prior theoretical advances such as Newtonian mechanics and Maxwell's electromagnetic theory are viewed from the standpoint of symmetry groups.

By far the deepest theoretical advance afforded by the group-theory approach is the set of ADE Coxeter graphs, which originally classified the most important (and useful) finite reflection groups, and in the form of Dynkin diagrams classified the most important (and useful) Lie groups. This work of the 1930s and 1940s has, in the last several decades evolved into the ADE classification of twenty-some mathematical categories, due to the work of many other mathematicians. The Russian mathematician, V. I. Arnold, as one of the most active and perceptive theoreticians in this process, wrote in his book, *Catastrophe Theory* [Arnold, 1986]:

"At first glance, functions, quivers, caustics, wave fronts and regular polyhedra have no connection with each other. But in fact, corresponding objects bear the same label not just by chance: for example, from the icosahedron one can construct the function $x^2 + y^3 + z^5$, and from it the diagram E_8, and also the caustic and wave front of the same name.

"To easily checked properties of one of a set of associated objects correspond properties of the others which need not be evident at all. Thus the relations between all the A, D, E-classifications can be used for the simultaneous study of all simple objects, in spite of the fact that the origin of many of these relations (for example, of the connections between functions and quivers) remains an unexplained manifestation of the mysterious unity of all things."

It is especially striking that the objects classified by the ADE graphs are of great utility in the advance of unified field theory afforded by superstring theory and its generalization to M-theory. The list of ADE classified categories (to be described in this book) should make this plain.

Lie algebras (and Lie groups): gauge group theory;

Kac–Moody (infinite-d) algebras;

Coxeter (reflection) groups, also called Weyl groups;

Coxeter arrangements;

Klein–DuVal singularities;

McKay correspondence groups (finite subgroups of $\mathcal{SU}(2)$);

Hyperspace crystallography;

Sphere-packing lattices (root lattices): error-correcting codes;

Quantizing lattices (weight lattices): analog/digital transforms;

Conformal field theories (living on the 2D string worldsheet);

Gravitational instantons (cf. Penrose twistors); ALE spaces;

Thom–Arnold catastrophe structures;

Heisenberg algebras (in various hyperspaces);

Generalized braid groups (cf. knots and links);

Quivers.

It is quite astonishing that this great diversity of categories should be classified by the simplest possible graphs.

Here is the complete list of ADE Coxeter graphs (and some salient correspondences) [Coxeter, 1973; McKay, 1980]:

Coxeter graph	McKay group	Coxeter number → Lie group dimension
A_n o-o-...-o (n nodes)	$\mathcal{Z}_{n+1}$	$n+1 \to n^2+2_n$

D_n o-o-...-o
    |
    o (n nodes) $n > 3$
	$\mathcal{Q}_{n-2}$	$2n-2 \to 2n^2-n$

E_6 o-o-o-o-o
      |
      o
	$\mathcal{TD}(24)$	$12 \to 78$

E_7 o-o-o-o-o-o
      |
      o
	$\mathcal{OD}(48)$	$18 \to 133$

E_8 o-o-o-o-o-o-o
      |
      o
	$\mathcal{ID}(120)$	$30 \to 248$

$\mathcal{TD}(24)$ means $\mathcal{TD}$ has 24 elements, etc.

Note that the Lie group dimension is $nk + n$, where n is the rank of the Coxeter graph, and k is the Coxeter number. This is just a taste of many relationships afforded by the structure of the ADE Coxeter graphs.

One striking feature of these Coxeter graphs is that, while there are the two infinities of A graphs and D graphs, there are only three E graphs: of ranks 6, 7 and 8. Moreover, these graphs correspond to the five Platonic solids:

$E_6 \to$ tetrahedron (self dual)

$E_7 \to$ octahedron (dual to the cube)

$E_8 \to$ icosahedron (dual to the dodecahedron).

Thus there are three Platonic symmetry groups corresponding to the three E-type graphs. It is plausible to think of the ADE graphs as the ultimate Platonic archetypes.

Let us call the study and applications of all the mathematical (and physical) categories classified by the ADE Coxeter graphs by the term ADEX theory, where X stands for yet to be discovered categories as well as possible extensions of the ADE graph system beyond the "simple" categories. As V. I. Arnold has shown, there is a vast beyond!

Chapter 2

The Octahedral Group

The most striking (and puzzling) aspect of the Standard Model of particle physics is the 3-family set of elementary particles, with two quarks and two leptons in each family. The three families carry the same charge assignments but increase in mass as we move down the family list. We can display this structural relationship in tabular form [Kaku, 1993; Veltman, 2003]:

Quarks	Leptons
Up	Electron neutrino
Down	Electron
Charm	Muon neutrino
Strange	Muon
Top	Tau neutrino
Bottom	Tau

These particles all have an intrinsic quantum spin of $1/2$ so they are all fermions (which are particles of $1/2$ integral spin). The quarks are fundamental hadrons, which means that they carry both electric charge and strong color charge. The quark electric charges are in units of $-1/3$ and $+2/3$, and the color charges are in units of three (conventional) colors. The quarks also carry weak charge.

Among the leptons, the electron, muon and tauon carry 1 unit of electric charge, while the neutrinos have no electric charge. All the leptons carry weak charge, but no strong (color) charge.

The basic rule of particle interaction is that fermions interact with each other by exchanging bosons (particles of integral spin). Also bosons can interact with other bosons by exchanging bosons.

The fundamental bosons are called gauge particles which correspond to the structure of certain Lie groups called gauge groups. The number of gauge particles corresponds to the dimensions of the gauge groups as follows:

Electrical interaction:	1 photon:	$\mathcal{U}(1)$ gauge group
Weak interaction:	3 weakons:	$\mathcal{SU}(2)$ gauge group
Strong interaction:	8 gluons:	$\mathcal{SU}(3)$ gauge group

In order to find some unifying structure in which these particles play their appropriate roles, we must first of all take into account these simple numbers: 6, 6, 1, 3, 8. Thus we have 6 quarks and 6 leptons in 3 families of 2 quarks and 2 leptons in each family. Also we have 1 photon, 3 weakons, and 8 gluons to account for.

Is there a mathematical structure that incorporates these quantities while also providing for the fermion and boson interaction rules? One might hope that a finite group would entail these structural rules.

Wonderfully, there is a very compact template for these structures. Let us look at the multiplication table of the symmetric-4 group, also called the octahedral group. This is the group of all 24 permutations of four objects. (Note: $24 = 4! = 4 \times 3 \times 2 \times 1$.) [Sirag, 1982; 1989; 1993; 1996].

There are several things to notice about this form of the octahedral group table.

(1) The fermions (quarks and leptons) carry flavor labels. The color assignments will correspond to the gauge group's fundamental representations, which will be embedded in the octahedral group algebra $\mathbb{C}[\mathcal{O}]$.

Multiplication table for $\mathcal{O}$ (i.e. $\mathcal{S}_4$):						Permutation	Particle	Coset
ABCD	**EFGH**	**IJKL**	**MNPQ**	**RSTV**	**WXYZ**	**Even:**	**Bosons:**	
ABCD	*EFGH*	*IJKL*	*MNPQ*	*RSTV*	*WXYZ*	**A** (1)(2)(3)(4)	Identon	1st
BADC	*FEHG*	*JIKL*	*PQMN*	*TVRS*	*XWZY*	**B** (12)(34)	Kleinon	(K_4)
CDAB	*GHEF*	*KLIJ*	*MNQP*	*VTSR*	*YZWX*	**C** (13)(24)	Kleinon	
DCBA	*HGFE*	*LKJI*	*QPNM*	*SRVT*	*ZYXW*	**D** (14)(23)	Kleinon	
EGHF	*IKLJ*	*ACDB*	*WZYX*	*MPNQ*	*RVST*	**E** (124)(3)	Familon	2nd
FHGE	*JLKI*	*BDCA*	*XYZW*	*PMQN*	*TSVR*	**F** (234)(1)	Familon	
GEFH	*KIJL*	*CABD*	*YXWZ*	*NQMP*	*VRTS*	**G** (143)(2)	Familon	
HFEG	*LJIK*	*DBAC*	*ZWXY*	*QNPM*	*STRV*	**H** (132)(1)	Familon	
ILJK	*ADBC*	*EHFG*	*RTSV*	*WYZX*	*MQPN*	**I** (142)(3)	Familon	3rd
JKIL	*BCAD*	*FGEH*	*TRVS*	*XZVW*	*PNMQ*	**J** (134)(2)	Familon	
KJLI	*CBDA*	*GFHE*	*VSTR*	*YWXZ*	*NPQM*	**K** (123)(4)	Familon	
LIKJ	*DACB*	*HEGF*	*SVRT*	*ZXWY*	*QMNP*	**L** (243)(1)	Familon	
						Odd:	**Fermions:**	
MQNP	*RSVT*	*WZYX*	*ACDB*	*EFHG*	*ILKJ*	**M** (24)(1)(3)	Quark	4th
NPMQ	*VTRS*	*YXWZ*	*CABD*	*GHFE*	*KJIL*	**N** (13)(2)(4)	Quark	
PNQM	*TVSR*	*XYZW*	*BDCA*	*FEGH*	*JKLI*	**P** (1234)	Lepton	
QMPN	*SRVT*	*ZWXY*	*DBAC*	*HGEF*	*LIJK*	**Q** (1432)	Lepton	
RVTS	*WYXZ*	*MNPQ*	*IJKL*	*ADCB*	*EGFH*	**R** (14)(2)(3)	Quark	5th
STVR	*ZXYW*	*QPNM*	*KLJI*	*DABC*	*HFGE*	**S** (23)(1)(4)	Quark	
TSRV	*XZWY*	*PQMN*	*JILK*	*BCDA*	*FHEG*	**T** (1342)	Lepton	
VRST	*YWZX*	*MNQP*	*KLIJ*	*CBAD*	*GEHF*	**V** (1243)	Lepton	
WXZY	*MPQN*	*RTSV*	*EHFG*	*IKJL*	*ABDC*	**W** (12)(3)(4)	Quark	6th
XWYZ	*PMNQ*	*TRVS*	*FGEH*	*JLIK*	*BACD*	**X** (34)(1)(2)	Quark	
YZXW	*NQPM*	*VSTR*	*GFHE*	*KILJ*	*CDBA*	**Y** (1423)	Lepton	
ZYWX	*QNMP*	*SVRT*	*HEGF*	*LJKI*	*DCAB*	**Z** (1324)	Lepton	

(2) The order of elements in this table is designed to make clear the 6 cosets of the Klein-4 group $\mathcal{K}_4$. This ordering is made possible by the fact that the octahedral group $\mathcal{O}$ has the alternating group $\mathcal{A}_4$ (= the tetrahedral group, $\mathcal{T}$) as a normal subgroup, and $\mathcal{T}$ has $\mathcal{K}_4$ as its normal subgroup, and $\mathcal{K}_4$ has $\mathcal{Z}_2$ as its normal subgroup.

Note that $\mathcal{T}$ is $\{A$ through $L\}$; $\mathcal{K}_4$ is $\{A, B, C, D\}$; $\mathcal{Z}_2$ is $\{A, B\}$; and A is the identity element. This normal subgroup series:

$$\mathcal{O} \to \mathcal{T} \to \mathcal{K}_4 \to \mathcal{Z}_2 \to \text{ Identity element } A$$

implies (according to Evariste Galois in 1831) that 4th degree polynomials have a general solution. Galois proved that 5th degree polynomials have no general solution, because $\mathcal{S}_5$ does not have such a normal series, due to the fact that the alternating group $\mathcal{A}_5$ is a simple group — thus having no normal subgroup. [Birkhoff and MacLane, 1965].

Since Galois explicitly used the octahedral group ($\mathcal{S}_4$) as the symmetry group on the four roots of 4th degree polynomials, it is plausible to conjecture that there is a deep connection between the particle structure implicit in the $\mathcal{S}_4$ table (as displayed here) and the fact that 4th degree polynomials are in general solvable (*a la* Galois).

(3) The cosets of $\mathcal{K}_4$ are displayed as separated into six sets ($\mathcal{K}_4$ itself being the identity coset). To generate left cosets of any subgroup $\mathcal{S}$ of a group $\mathcal{G}$, one multiplies each element from the left:

$$xz, \quad \text{where } x \text{ is any element of } \mathcal{G}, \quad \text{and} \quad z \in \mathcal{S}.$$

Similarly, right cosets are generated by multiplication from the right: zx.

A normal subgroup has equivalent left and right cosets within the group. Since $\mathcal{K}_4$ is a normal subgroup of $\mathcal{T}$, the fact that the left and right cosets are equivalent allows the $\mathcal{S}_4$ table to be displayed as above. Thus the 36 coset patches make up the (6×6) multiplication table within the (24×24) table.

(4) The 36 coset patches can themselves be regarded as the symmetric-3 group, consisting of the six permutations of three objects). This corresponds to the fact that $\mathcal{S}_4 = \mathcal{K}_4 \times \mathcal{S}_3$.

(5) Interweaving the $\mathcal{K}_4$ coset structure in $\mathcal{S}_4$ is the conjugacy class structure of $\mathcal{O}$. A conjugacy class within a group is a similarity set. We say that h and h' are conjugate if

$$h' = ghg^{-1}$$

(where g^{-1} is the inverse of g) for some element g in the group. Thus we call ghg^{-1} a similarity transformation of h.

For any symmetric group $\mathcal{S}_n$ (which is the set of all permutations of n objects) the classes are equivalent to the permutation types within $\mathcal{S}_n$. The five-permutation types are indicated by the cycle patterns listed with the elements of $\mathcal{S}_4$. The cardinalities of these five-cycle patterns fit exactly the cardinalities of the fundamental fermions (6 quarks and 6 leptons) and the fundamental gauge bosons (1 photon, 3 weakons, and 8 gluons). The bosonic particles correspond to the even permutations, while the fermionic particles correspond to the odd permutations. A permutation can be seen as a series of swaps between pairs of objects within the permutation set. Even permutations are equivalent to an even number of swaps, while odd permutations are equivalent to an odd number of swaps.

Note that for permutations: even × even = even; even × odd = odd; odd × even = odd; odd × odd = even. This pattern is depicted in the $\mathcal{S}_4$ table by the substructure corresponding to the $\mathcal{Z}_2$ group table, which exactly matches the even–odd multiplication pattern.

This is significant, since the basic rule of particle interactions (mentioned above) is that fermions interact by exchanging bosons.

Thus one can interpret the $\mathcal{S}_4$ permutations to correspond to simple Feynman diagrams, of which there are two types. Two fermion lines meet at a vertex with an interaction boson line. Two boson lines meet at a vertex with an interaction boson line.

(6) Although we can interpret the octahedral group ($\mathcal{S}_4$) table as mapping particle interactions, these interactions do not exactly track the Standard Model interactions. For this reason, I have given the gauge boson classes different labels, which more correctly characterizes their interaction behavior. For example the "familons" entail transformations between the three families of quarks and leptons. But notice that this transformation seems to be a substructure of Standard Model interaction. It would seem that there is a mapping of the Standard Model interactions to a more basic set of interactions. We can think of these interactions as being due to the permutations of four basic entities, which I have previously called zons. They can be quite abstract entities.

These octahedral zon permutation classes track perfectly the cardinalities of the Standard Model classes: {1, 3, 8} for gauge bosons; and {6, 6} for quarks and leptons.

Feynman diagrams display two types of vertices. The first type corresponds to the particle rule: fermions interact with fermions by exchanging a boson. The second type of vertex shows up in the weak interaction because weakons can interact with each other. These two types of interaction are depicted here:

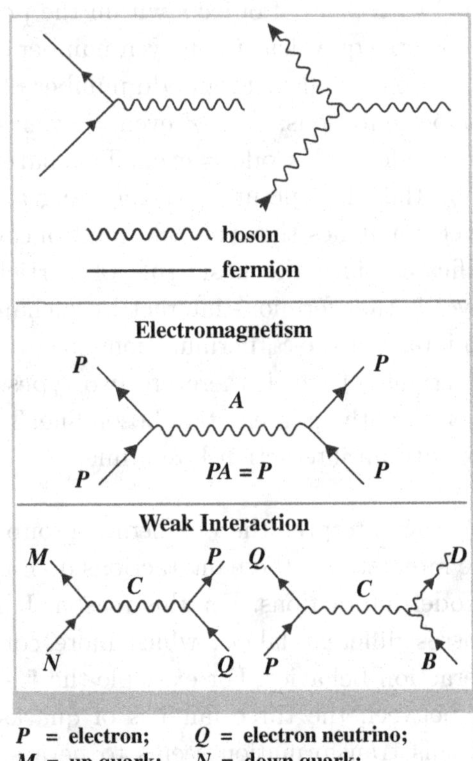

Fig. 1. Feynman diagrams.

Actually A is somewhat different from a photon, so I call it an identon, because it corresponds to the identity element of the group. Also B, C and D are somewhat different from the weakons, so I call them Kleinons after Felix Klein, for whom the Klein-4 group $\mathcal{K}_4$ (consisting of A, B, C and D) is named [Klein, 1956].

I will recover the standard gauge theory of these force particles by embedding the gauge groups $\mathcal{U}(1) \times \mathcal{SU}(2) \times \mathcal{SU}(3)$ in the octahedral group-algebra $\mathbb{C}[\mathcal{O}]$.

Note that the letters on these Feynman-type diagrams have been taken directly from the octahedral multiplication table. Thus reading from the table we find the relevant items:

For the electromagnetic interaction: $PA = P$

$P =$ electron
$A =$ identon ($\sim$ photon)

For the weak interaction:

$NC = M$
$QC = P$
$PC = Q$
$CB = D$

$Q =$ electron neutrino
$P =$ electron
$M =$ up-quark
$N =$ down-quark
$C =$ kleinon ($\sim$ weakon)
$B =$ kleinon ($\sim$ weakon)
$D =$ kleinon ($\sim$ weakon)

The next Feynman-type diagram will model the proton (udu) to neutron (udd) via the exchange of a Pi meson ($\underline{\text{d}}$u), where $\underline{\text{d}}$ is the anti-down quark.

Note that in this diagram, all lines meeting at a vertex are modeled by the multiplications indicated by the octahedral ($\mathcal{S}_4$) group table.

It has frequently been noticed that the 3-fermion family structure must be a hint of particle substructure [Lincoln, 2013]. However, the zon substructure proposed here is more radical since the $\mathcal{S}_4$ substructure underlies both the 3-fermion families, as well as the three Standard Model gauge bosons.

Thus there must be a transformation that maps the octahedral permutation group to the Standard Model gauge groups and their actions on the fundamental Standard Model fermions. We now turn to this transformation structure.

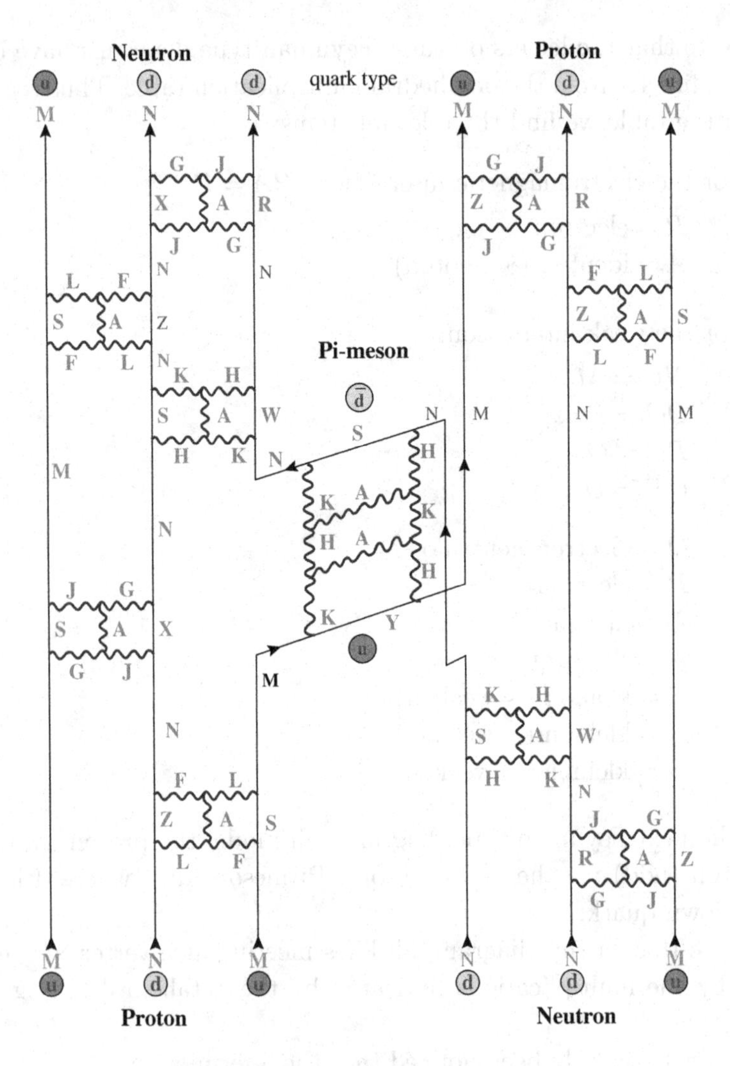

Fig. 2. Neutron to proton interaction.

The transformation mapping the octahedral group to the Standard Model gauge groups is accomplished by the mathematical transformation: $\mathcal{O} \rightarrow \mathbb{C}[\mathbb{O}]$, where $\mathbb{C}[\mathcal{O}]$ is a complex 24D vector space, with O as basis. This transformation is actually quite intricate and formally depends on the representation structure of the finite group.

For a symmetric group, such as $\mathcal{O}$, this is accomplished by way of the character table for $\mathcal{O}$.

The character table for symmetric-4 (octahedral) group.

Class cardinality:	{1}	{3}	{8}	{6}	{6}
Class cycle pattern:	(1111)	(22)	(31)	(211)	(4)
Representation (iirep):					
[4]	1	1	1	1	1
[22]	2	2	−1	0	0
[211]	3	−1	0	−1	1
[31]	3	−1	0	1	−1
[1111]	1	1	1	−1	−1

The dimensions of the five basic representations are indicated by the numbers in the first column (the identity element column). These numbers are called characters and are traces of the representation matrices. Since a trace is the sum of the diagonal elements, the trace of the identity element (whose matrix has 1's along the prime diagonal and zeros elsewhere) reveals the dimension of the representation matrix.

The Class permutation cycle patterns correspond to the representation structure by way of the Young diagrams [James and Kerber, 1981]:

Young diagrams for S_4 derived from S_4 classes.

Iirep Dim.:	1	2	3	3	1
Cycle Pattern:	[4]:	[22]:	[211]:	[31]:	[1111]:
Young Diag.:	◇◇◇◇	◇◇ ◇◇	◇◇ ◇ ◇	◇◇◇ ◇	◇ ◇ ◇ ◇

These Young diagrams are used to generate the five inequivalent-irreducible representations (iireps) of the symmetric-4 group. These representations are basic in the sense that any representation will be a direct product of these iireps.

Notice that the Young patterns correspond directly to the Class cycle patterns. In general, for any finite group the number of classes corresponds exactly to the number of iireps. However, for any symmetric-n group (the set of all permutations of n objects), the correspondence is more straightforward.

Once we have the five sets of representation matrices (as generated by standard procedures from the Young diagrams), we can use the $\mathcal{O}$ group basis to generate the matric basis of the group algebra $\mathbb{C}[\mathcal{O}]$. The transformation formula is rather intricate, and will require some unpacking in order to make it clear. Here is the group-basis-to-matric-basis transform:

$$(f^a/g)\sum_x \{[(M^{-1})_x]_{sr}\}^a G_x = [e_{rs}]^a,$$

where g is the number of elements in the group, f^a is the degree of the a-th representation; and $\{[(M^{-1})_x]_{rs}\}^a$ is the (s,r)-th element in the $f^a \times f^a$ matrix $[(M^{-1})_x]^a$, which represents the a-th irreducible representation of the group element $(G_x)^{-1}$. Then $[e_{rs}]^a$ is the number 1 as the (r,s)-th element in the a-th simple matrix algebra basis. Note that a ranges over the number of classes.

Thus the full set of matrices of various size, with 1 in the (r,s)-th position and zeros elsewhere, form the matric basis of the algebra [Matsen, 1975].

Each total matric algebra of dimension-d consists of all $d \times d$ matrices. Since $\mathbb{C}[\mathcal{O}]$ is a complex group algebra, these matric algebras consist of complex matrices. Thus $\mathbb{C}[\mathcal{O}]$ as a matrix algebra is equivalent to the direct sum of the five total matrix algebras:

$$M^{\mathbb{C}}(1) \oplus M^{\mathbb{C}}(2) \oplus M^{\mathbb{C}}(3) \oplus M^{\mathbb{C}}(3) \oplus M^{C}(1).$$

Note that the sum of the squares of the dimensions $\{1, 2, 3, 3, 1\}$ is $1+4+9+9+1$, which is 24. Thus the five matric algebras span the 24 dimensions of $\mathbb{C}[\mathbb{O}]$.

The Standard Model gauge groups are unitary groups, so it is useful that we have here a sum of complex matric algebras. This is because every complex d-dimensional matrix embeds a d-dimensional unitary matrix, which are elements of a unitary group. So we can simply write:

$$\mathbb{C}[\mathcal{O}] \to \mathcal{U}(1) \times \mathcal{U}(2) \times \mathcal{U}(3) \times \mathcal{U}(3) \times \mathcal{U}(1),$$

where the arrow indicates a projection to the embedded unitary group.

The unitary group (consisting of the direct product of five unitary groups) can also be thought of as the set of all the unitary elements in the complex octahedral group algebra $\mathbb{C}[\mathcal{O}]$. And we note that we could have written this unitary group product merely by looking at the dimensions of the octahedral group iireps in the octahedral character table. But the reason that this works out is actually because of the complicated summing over all the octahedral elements in the formula, transforming the group algebra $\mathbb{C}[\mathcal{O}]$ from a group-basis to a matric-basis.

A d-dimensional unitary element acts on a complex d-dimensional vector space and rotates the vectors without stretching or shrinking. This is the complex dimensional analog to orthogonal group actions on real vector spaces.

Now since $\mathcal{U}(n) = \mathcal{U}(1) \times \mathcal{SU}(n)$, we can rearrange the structure of the product of the five unitary groups to become:

$$\mathbb{C}[\mathcal{O}] \to \mathcal{T}^4 \times \mathcal{U}(1) \times \mathcal{SU}(2) \times \mathcal{SU}(3) \times \mathcal{SU}(3).$$

Thus we see immediately that we have the Standard Model set of gauge groups, $\mathcal{U}(1) \times \mathcal{SU}(2) \times \mathcal{SU}(3)$, along with an extra $\mathcal{SU}(3)$, as well as a four-dimensional torus, $\mathcal{T}^4$, consisting of the product of four $\mathcal{U}(1)$ groups.

We can organize this structure as a principal fiber bundle, with $\mathcal{T}^4$ as the base space and $\mathcal{U}(1) \times \mathcal{SU}(2) \times \mathcal{SU}(3) \times \mathcal{SU}(3)$ as fiber. This is the mathematical nomenclature for the Standard Model gauge theory. There is a dictionary transforming fiber bundle terms to gauge physics terms [Schutz,1980, Bleecker, 1981].

Base Space $\leftrightarrow$ Spacetime

Fiber $\leftrightarrow$ Gauge Group

Local Section $\leftrightarrow$ Coordinate Basis

Change of Local Section $\leftrightarrow$ Gauge Transformation

The geometrical picture here is that attached to each point of the base space (spacetime) is an internal space, called a fiber. The word "fiber" suggests a one-dimensional object, but these fibers are of arbitrary dimension. In the case of the Standard Model principal bundle, the fiber is $\mathcal{U}(1) \times \mathcal{SU}(2) \times \mathcal{SU}(3)$, which is a 12-dimensional space, having both geometric and algebraic properties. In my version, there is an extra $\mathcal{SU}(3)$, making the fiber into a 20-dimensional space.

Moreover, the base space is not the standard Minkowski spacetime, but rather the 4-torus $\mathcal{T}^4$. This picture will be modified later, but for now we can consider $\mathcal{T}^4$ as a "toy model" for a 4D base space, pending a more realistic formalism.

The fiber bundle picture incorporates the gauge particles, which are generators of the gauge groups (i.e. the principal fiber). But it also incorporates the basic matter particles. This is accomplished by the set of vector spaces on which the gauge groups act. This is mathematically formalized as a set of vector bundles associated with the principal bundle. It is this association which justifies the use of the term "principal bundle." In particle physics theory, these vector spaces (which correspond to the fibers of the associated vector bundles) are parametrized by the various basic particle states.

By embedding the Standard-Model gauge groups in a larger gauge group, it is possible to embed their associated vector spaces in larger vector spaces. This is the strategy of Grand Unified Theory, such as the $\mathcal{SU}(5)$ theory of [Georgi and Glashow, 1974]. In string theory, which also incorporates quantum gravity, $\mathcal{SU}(5)$ is embedded in much higher dimensional gauge groups.

At this point, we are encouraged by the fact that starting from the (rather unique) structure of the octahedral group as a template for fundamental particle interactions, we have derived a reasonable facsimile of the Standard Model principal fiber bundle embedded in the group algebra $\mathbb{C}[\mathcal{O}]$.

Chapter 3

The Octahedral Double Group

In order to include gravity gauge groups, it is necessary to go beyond the embedding of the Standard Model groups, $\mathcal{U}(1)\times\mathcal{SU}(2)\times\mathcal{SU}(3)$, in a space that is higher dimensional than the 24-dimensional group algebra $\mathbb{C}[\mathcal{O}]$. We will take this embedding one step at a time.

The first step is to enlarge the group from the octahedral group $\mathcal{O}$ to the octahedral double group $\mathcal{OD}$. In the mathematical literature, following [Coxeter, 1973], this group is called the binary octahedral group. Here I will be using the double group nomenclature of quantum chemistry [Chisholm, 1976; Flurry, 1980].

The octahedral group $\mathcal{O}$ is a finite subgroup of $\mathcal{SO}(3)$, the rotation group on the three-dimensional real vector space $\mathcal{R}^3$. However, $\mathcal{OD}$ is a finite subgroup of $\mathcal{SU}(2)$, the rotation group on the two-dimensional complex vector space. Thus, since $\mathcal{SU}(2)$ is the double cover of $\mathcal{SO}(3)$, $\mathcal{OD}$ is the double cover of $\mathcal{O}$. We can write:

$$\mathcal{SU}(2)/\{\pm 1\} = \mathcal{SO}(3); \quad \mathcal{OD}/\{\pm 1\} = \mathcal{O}.$$

Note that $\mathcal{O}$ is not a subgroup of $\mathcal{OD}$, but is rather a quotient group. The divisor $\{\pm 1\}$ is isomorphic to $\mathcal{Z}_2$, the group of integers, mod 2, and is also called the cylic-2 group.

The group $\mathcal{SU}(2)$ is, by definition, the set of all special-unitary transformations of $\mathbb{C}^2$, and its basic representation is equivalent to the set of special-unitary 2×2 complex matrices.

For a unitary transformation $\mathcal{U}$, the inverse $\mathcal{U}^{-1}$ is equivalent to the complex-conjugate transpose $\mathcal{U}^*$, and thus

$$\mathcal{U}\mathcal{U}^* = 1.$$

A unitary matrix has its determinant equal to ± 1, while a special-unitary matrix has its determinant equal to $+1$.

We can note in passing that $\mathcal{SU}(2)$ is the simple compact Lie group whose Coxeter label is A_1, which corresponds to the Coxeter diagram consisting of a single node.

As we will see later, $\mathcal{OD}$ corresponds to the Coxeter label (and diagram) E_7, by way of the McKay correspondence, which plays a prominent role in this book.

The character table for the $\mathcal{OD}$ group (8 classes, $\mathcal{C}_i$, and 8 iireps, $\mathcal{R}_i$).

		$1\mathcal{C}_0$	$6\mathcal{C}_1$	$8\mathcal{C}_2$	$12\mathcal{C}_3$	$6\mathcal{C}_4$	$1\mathcal{C}_5$	$8\mathcal{C}_6$	$6\mathcal{C}_7$
	$\mathcal{R}_0$	1	1	1	1	1	1	1	1
$\rightarrow$	$\mathcal{R}_1$	2	0	1	0	$\sqrt{2}$	-2	-1	$-\sqrt{2}$
	$\mathcal{R}_2$	3	-1	0	-1	1	3	0	1
$\rightarrow$	$\mathcal{R}_3$	4	0	-1	0	0	-4	1	0
	$\mathcal{R}_4$	2	2	-1	0	0	2	-1	0
	$\mathcal{R}_5$	3	-1	0	1	-1	3	0	-1
$\rightarrow$	$\mathcal{R}_6$	2	0	1	0	$-\sqrt{2}$	-2	-1	$\sqrt{2}$
	$\mathcal{R}_7$	1	1	1	-1	-1	1	1	-1

Note that the number of elements in each class is indicated along with the class label. These numbers add up to the order of the $\mathcal{OD}$ group: 48, which is also the sum of the squares of the iirep dimensions: $1 + 4 + 9 + 16 + 4 + 9 + 4 + 1 = 48$. The arrows indicate the three representations beyond the 5 of the $\mathcal{O}$ group.

The ordering of the iireps is designed to accord with the indexing of the extended E_7 Coxeter graph, which will be employed later.

$\underline{0}\ \ \underline{1}\ \ \underline{2}\ \ \underline{3}\ \ \underline{5}\ \ \underline{6}\ \ \underline{7}$

1–2–3–4–3–2–1

|

2 $\underline{4}$

The numbers labeling this extended graph correspond to the $\mathcal{OD}$ iirep dimensions, which are indicated by the first column of the $\mathcal{OD}$ character table. (The indices are the underlined numbers.) Each character is the trace of the iireps for each class. Since the first column is the identity-element column, the trace reveals directly the dimension of the corresponding iirep.

In Chap. 2, we progressed from the octahedral group to the octahedral group-algebra: $\mathcal{O} \to \mathbb{C}[\mathcal{O}]$. Now we will progress from the octahedral double group to the octahedral double group-algebra.

$$\mathcal{OD} \to \mathbb{C}[\mathcal{OD}].$$

Previously we found that the group algebra $\mathbb{C}[\mathcal{O}]$ consisted of the direct sum of five total matrix algebras:

$$M^{\mathbb{C}}(1) \oplus M^{\mathbb{C}}(2) \oplus M^{\mathbb{C}}(3) \oplus M^{\mathbb{C}}(3) \oplus M^{\mathbb{C}}(1).$$

Similarly, we find that the group algebra $\mathbb{C}[\mathcal{OD}]$ consists of the direct sum of eight total matrix algebras:

$$M^{\mathbb{C}}(1) \oplus M^{\mathbb{C}}(2) \oplus M^{\mathbb{C}}(3) \oplus M^{\mathbb{C}}(3) \oplus M^{\mathbb{C}}(1) \oplus M^{\mathbb{C}}(2) \oplus M^{\mathbb{C}}(2) \oplus M^{\mathbb{C}}(4).$$

Since the first five terms in this sum is the octahedral group algebra, we see immediately that:

$$\mathbb{C}[\mathcal{OD}] = \mathbb{C}[\mathcal{O}] \oplus M^{\mathbb{C}}(2) \oplus M^{\mathbb{C}}(2) \oplus M^{\mathbb{C}}(4).$$

Then we can rewrite this direct sum in terms of Clifford algebras:

$$\mathbb{C}[\mathcal{OD}] = \mathbb{C}[\mathcal{O}] \oplus P^{\mathbb{C}} \oplus D^{\mathbb{C}},$$

where $P^{\mathbb{C}}$ is the complex Pauli algebra equivalent to $Cl(3) \otimes \mathbb{C}$, and $D^{\mathbb{C}}$ is the complex Dirac algebra equivalent to $Cl(4) \otimes \mathbb{C}$ [Bleecker, 1981; Porteous, 1981].

We see immediately that $P^{\mathbb{C}}$ provides for the spin structure of particle physics, while $D^{\mathbb{C}}$ provides for the anti-matter structures of particle physics.

Moreover, since $M^{\mathbb{C}}(4)$ is the set of all 4×4 complex matrices, we will find embedded in the complex matrix algebra the Lie groups:

$\mathcal{SO}(1,3)$, which is a quotient group of $\mathcal{SU}(2,2)$.

Note that $\mathcal{SO}(1,3)$ is the Lorentz group providing the symmetry structure for Minkowski space in special relativity.

Also, $\mathcal{SU}(2,2)$ is more closely related to various general relativity symmetry structures [Ward and Wells, 1990].

It should also be noted that $\mathcal{OD}$, as a finite subgroup of $\mathcal{SU}(2)$, has a representation as 48 quaternions, with 24 positive and 24 negative elements. The positive elements will be symbolized by lower-case letters matching the upper-case elements of $\mathcal{O}$:

$$a \to \begin{vmatrix} 1 & 0 \\ 0 & 1 \end{vmatrix} \qquad b \to \begin{vmatrix} 0 & -i \\ -i & 0 \end{vmatrix} \qquad c \to \begin{vmatrix} 0 & -1 \\ 1 & 0 \end{vmatrix} \qquad d \to \begin{vmatrix} -i & 0 \\ 0 & i \end{vmatrix}$$

where i is $\sqrt{-1}$, so that: $a = -i\sigma_x, b = -i\sigma_y, c = -i\sigma_z$, and where σ symbolizes the standard Pauli spin-matrices. Rewriting Hamilton's famous quaternion formula in terms of a, b, c and d, we have:

$$b^2 = c^2 = d^2 = bcd = -a.$$

The remaining eight Bosonic elements of $\mathcal{O}$, i.e. the Familons, E through L, correspond to the $\mathcal{OD}$ quaternions:

$$\begin{array}{ll} e \to \frac{1}{2}(a - b + c - d) & i \to \frac{1}{2}(a + b - c + d) \\ f \to \frac{1}{2}(a + b + c + d) & j \to \frac{1}{2}(a - b + c + d) \\ g \to \frac{1}{2}(a + b - c - d) & k \to \frac{1}{2}(a + b + c - d) \\ h \to \frac{1}{2}(a - b - c + d) & l \to \frac{1}{2}(a - b - c - d) \end{array}$$

Elements of $\mathcal{OD}$ corresponding to the first fermion family of $\mathcal{O}$:

$$\begin{array}{ll} m \to \frac{1}{\sqrt{2}}(d - b) & p \to \frac{1}{\sqrt{2}}(a - c) \\ n \to \frac{1}{\sqrt{2}}(d + b) & q \to \frac{1}{\sqrt{2}}(a + c) \end{array}$$

Elements of $\mathcal{OD}$ corresponding to the second fermion family of $\mathcal{O}$:

$$\begin{array}{ll} r \to \frac{1}{\sqrt{2}}(b + c) & t \to \frac{1}{\sqrt{2}}(a - d) \\ s \to \frac{1}{\sqrt{2}}(b - c) & v \to \frac{1}{\sqrt{2}}(a + d) \end{array}$$

Elements of $\mathcal{OD}$ corresponding to the third fermion family of $\mathcal{O}$:

$$\begin{array}{ll} w \to \frac{1}{\sqrt{2}}(c + d) & y \to \frac{1}{\sqrt{2}}(a - b) \\ x \to \frac{1}{\sqrt{2}}(c - d) & z \to \frac{1}{\sqrt{2}}(a + b) \end{array}$$

The 24 negative elements of $\mathcal{OD}$ are generated by multiplying each element of the 24 positive elements by $-a$, which is represented as -1. For example, the negative of e is $-e$, and is transformed from:

$$\frac{1}{2}(1 - a + b - c) \quad \text{to} \quad \frac{1}{2}(-1 + a - b + c)$$

while the negative of m is $-m$, and is transformed from:

$$\frac{1}{\sqrt{2}}(c - a) \quad \text{to} \quad \frac{1}{\sqrt{2}}(a - c).$$

Note also that the eight elements $\{a, b, c, d, -a, -b, -c, -d\}$ make up the finite quaternion group, which is the double cover of the 4-element Klein group $\mathcal{K}_4$. The quaternion group table can be symbolized by the triangle:

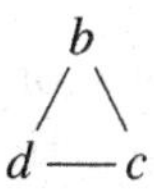

which can be read in a clockwise direction to mean:

$$bc = d, \quad cd = b, \quad db = c$$

while in the counterclockwise direction:

$$bd = -c, \quad dc = -b, \quad cb = -d.$$

The identiy element, a, and its negative, $-a$, function as ordinary multiplication by 1 and -1.

We can see that quaternionic multiplication is not commutative, since for example:

$$bc = d \quad \text{while} \quad cb = -d; \quad bcd = -1$$

as required by Hamilton's original formula in terms of $\{i, j, k\}$. Here we have avoided this common convention, because we have other uses for the symbols i, j and k.

An important question arises at this point. From the particle point of view, what is the role of all 48 elements of $\mathcal{OD}$, as the double cover of $\mathcal{O}$?

I would conjecture that this doubling of the particle labels corresponds to the doubling of supersymmetry (informally called SUSY) [Wess and Bagger, 1983; Kane, 2000].

There would be a parallelism between this SUSY doubling and the doubling of particles via the antimatter transformation called "charge conjugation." This parallelism between the antimatter and SUSY transformations is quite striking for the two extensions of quantum theory:

Q. Th. + Special relativity requires antimatter partners.

Q. Th. + General relativity requires SUSY partners.

Remember that both $\mathcal{O}$ and $\mathcal{OD}$ are templates for the particle states. This is a level of structure which is dual to the gauge group (and spacetime) structure implicit in the transformation from:

$$\mathcal{OD} \to \mathbb{C}[\mathcal{OD}] = \mathbb{C}[\mathcal{O}] \oplus \mathcal{P}^{\mathbb{C}} \oplus \mathcal{D}^{\mathbb{C}}.$$

Moreover, the representations of the gauge groups correspond to Standard Model versions of particle states. These particle states are ordinarily said to reside in an "internal" space. Mathematically speaking, however, the particle states are to be found in the vector bundles associated with the principle fiber bundle, whose base space is spacetime. Each representation corresponds to a different vector bundle.

The gauge particles are special in the sense that they are represented as tangent vectors in the tangent bundle, which is the direct product of the gauge group and the Lie algebra that generates the gauge group. Indeed, the set of unit-length tangent vectors at the identity element of the gauge group constitutes the Lie algebra of the gauge group (which is a Lie group).

The representation of a Lie group by its action on its own Lie algebra is called the adjoint representation. And this accounts for the fact that the number of gauge particles corresponding to a gauge group matches the dimensionality of the gauge group.

We can postulate that the doubling of particle states via the transformation:

$$\mathcal{O} \to \mathcal{OD}$$

is a template for the supersymmetry doubling of particle states. Given this postulate, we expect to see a relationship between $\mathcal{OD}$ and the algebraic structures of supersymmetry, supergravity, and superstring theory.

This relationship is engendered by the McKay correspondence which makes a duality between $\mathbb{C}[\mathcal{OD}]$ and the E_7 Lie algebra. We now turn to this McKay correspondence.

Chapter 4

The McKay Correspondence

John McKay [1980; 1981] published an astonishing correspondence between the ADE Lie algebras and the group algebras of the finite subgroups of $\mathcal{SU}(2)$. This correspondence is based on an exact equivalence between the ADE balance numbers and the dimensions of the iireps of the finite subgroups of $\mathcal{SU}(2)$. Since these balance numbers are labels on the graphs of the extended ADE graphs, this suggests a relationship between the affine Lie algebras and the finite subgroup of $\mathcal{SU}(2)$, which I will henceforth refer to as McKay groups.

The affine Lie algebras are the infinite-dimensional Lie algebras, whose largest finite-dimensional subalgebras are the ordinary simple Lie algebras. This subalgebra relationship is true for any of the simple Lie algebras, but here we will treat only the affine Lie algebras classified by the ADE Coxeter (Dynkin) graphs.

The term "affine" is based on Coxter's distinction between finite order reflection groups (which have a finite number of elements) and infinite order reflection groups. The finite reflection groups could be modeled by a central projection of the reflection hyperplanes to a hyperspherical space, creating spherical simplexes. The infinite reflection groups could be modeled only by mirrors in a flat (affine) space creating Euclidean simplexes [Coxeter, 1973].

The reflection mirrors are hyperplanes of dimension $n-1$ embedded in an n-dimensional real Euclidean space $\mathfrak{R}^n$, which I call a reflection space. There are n basic mirrors in $\mathfrak{R}^n$. Each mirror reflection transforms the vectors from one side of the mirror to the other without altering the length of each vector. Each mirror reflection leaves invariant its own internal vectors. However, mirrors are reflected through each other to create new virtual mirrors. This can be clearly pictured in the 2D reflection space corresponding to the A_2 Coxeter graph.

The A_2 Reflection Space

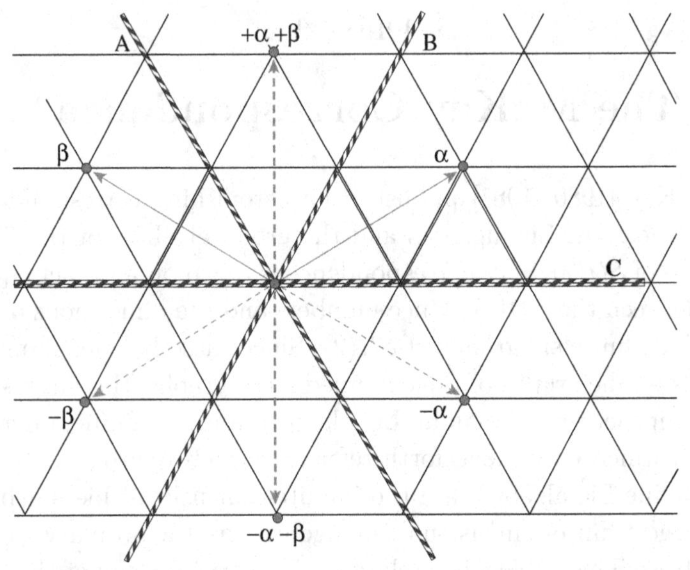

A and **B** are basic mirrors, set at 60° to one-another. **C** is a mirror which itself is generated by reflection in either **A** or **B**. α and β are **basic roots**, set at 120° to one-another. The other four roots are generated by reflections in the mirrors. The **nodes** (crossing points) are the weight lattice.
Note: The root lattice is a sub-lattice of the weight lattice.

It has long been known that there is a 1–1 correspondence between the simple Lie algebras of ADE type and the appropriate subset of Coxeter's reflection groups (also of ADE type). Both of these algebraic structures are classified (and modeled) by the singly laced Coxeter (Dynkin) graphs. These finite groups are called Coxeter (reflection) groups, and are also called Weyl groups associated with the corresponding Lie Algebras.

The affine Coxeter graphs are generated by extending the ordinary Coxeter graph by a single node. There is also a number associated with each node, which I call a balance number. This is because the extra node must correspond to a vector that balances all the basic mirror vectors (orthogonal to each mirror). There is one basic mirror corresponding to each node of the Coxeter graph and a mirror vector (called a basic root) attached to each mirror at the origin of the reflection space. This origin is the point where all the basic mirrors intersect.

The extended Coxeter graphs (with balance numbers) and the corresponding McKay groups are as follows:

```
A_n      1–1–...-1        Z_{n+1}
          \     /
           \   /
            1*

D_n     1-2-...-2-1       Q_{n-2}
          |     |
          1     1*

E_6     1–2–3–2–1         TD
            |
            2
            |
            1*

E_7   1*–2–3–4–3–2–1      OD
             |
             2

E_8   2–4–6–5–4–3–2–1*    ID
          |
          2
```

For each ADE graph, the balance numbers encode the lengths of the basic mirror vectors necessary to allow the added vector (indicated by 1*) to balance all the other mirror vectors. If we consider the basic mirror vectors as force vectors, the additional vector of length 1 (attached at the right place) would make the total force zero; and thus the set of force vectors would be in balance. We can formally write:

$$\sum x_i y^i = 0,$$

where x_i is a balance number, y^i is a basic mirror vector (of unit length); and the sum is over 1 through $n+1$, for an ADE graph of type $\mathcal{X}_n$ [Coxeter, 1973].

Historical Note: The reflection groups (both finite and infinite) are usually called Weyl groups because in 1935, at Princeton, Herman Weyl distributed his mimeographed lecture notes called "The structure and representations of continuous groups." However, the reflection groups were included in an appendix ("Discrete groups generated by reflections") by Coxeter, for whom they should be named. Moreover, in 1934 Coxeter had already published a paper with the same title in which he applied his graphs to the theory of simple Lie groups [Coxeter, 1934; Akivis and Rosenfeld, 1991, pp. 60–64].

Elie Cartan in 1928 had introduced the distinction between hyperspherically modeled reflection groups (of finite order) and the modeling of affine reflection groups (of infinite order) in $\mathfrak{R}^n$. Coxeter in 1934 introduced his graphs and proved that Cartan's classification of both finite and infinite reflection groups was complete [Coxeter, 1973].

It was McKay who first noticed that the balance numbers on the extended ADE Coxeter graphs (for affine reflection groups) matched exactly the iireps of the corresponding finite subgroups of $\mathcal{SU}(2)$.

Since these extended ADE graphs were already being used to classify ADE-type affine Lie algebras, McKay dug deeper into the correspondence between these two very different algebraic structures. He was able to prove the following propositions [McKay, 1980, 1981]:

McKay Proposition 1. *For every ADE Affine Lie algebra $\mathcal{X}$: the Cartan Matrix of $\mathcal{X}$ has eigenvectors which are the columns of the character table of the corresponding finite subgroups of $\mathcal{SU}(2)$.*

Note: This is why I call the finite subgroups of $\mathcal{SU}(2)$ "McKay groups."

The Cartan matrix $\mathcal{C}$ of a simple Lie algebra is a square matrix which encodes the data necessary to generate the Lie algebra. The dimension of the matrix is the dimensionality of the Cartan subalgebra $\mathfrak{h}$ of the Lie algebra $\mathcal{L}$, where $\mathfrak{h}$ is the largest commutative subalgebra of $\mathcal{L}$.

Starting from any ADE Coxeter graph, one can generate $\mathcal{C}$. Or starting from $\mathcal{C}$, one can generate the Coxeter graph. This is because:

$$\mathcal{C} = 2\mathcal{E} - \mathcal{A}$$

where $\mathcal{E}$ is the identity matrix (with 1's along the diagonal) and $\mathcal{A}$ is the adjacency matrix (whose component a_{ij} is 1 if the i-th node is adjacent to the j-th node; and otherwise is 0). The rows of $\mathcal{C}$ encode the coordinates of the basic roots, with respect to a basis of $\mathfrak{h}$.

See Appendix A, for the concrete example of the $\mathcal{OD}$ case.

From the basic roots, one can generate the entire Lie algebra, because these roots are vectors orthogonal to basic mirrors which generate a complete set of roots for the Lie algebra, whose total dimension is equal to the total number of roots plus the dimensionality of the Cartan subalgebra $\mathfrak{h}$. The basis elements of $\mathfrak{h}$ are labeled as H_i $\{i, \ldots, r\}$ where r is the rank of the Coxeter graph. Then the Lie algebra $\mathcal{L}$ can be described via the following structure equations:

$$\begin{aligned}
[H_i, H_j] &= 0, \\
[H_i, E_j] &= \mathcal{C}_{ij} E_j, \\
[H_i, F_j] &= -\mathcal{C}_{ij} F_j, \\
[E_i, F_j] &= \delta_{ij} H_i,
\end{aligned}$$

where $1 \leq i, j, \leq r$, and $[X, Y]$ is the Lie product $XY - YX$; and where δ_{ij} are the components of the unit matrix, with 1's on the diagonal and zeros elsewhere.

Note that in these formulas, E_j are eigenvectors of H_i with eigenvalue $\mathcal{C}_{ij}$. So that a set of r eigenvalues for E_j locate a basic positive root in $\mathfrak{h}$. Similarly, the formulas for F_i generate a basic negative root in $\mathfrak{h}$.

The remaining roots can be generated via the adjoint representation (symbolized by ad), consisting of matrices, whose dimensionality is equal to that $\mathfrak{L}$. The adjoint formulas are:

$$(\text{ad } E_i)^{-\mathcal{C}_{ij}+1} E_j = 0, (\text{ad } F_i)^{-\mathcal{C}_{ij}+1} F_j = 0, \qquad \text{for } i \neq j,$$

where $(\text{ad } x)y = [x, y] = xy - yx$. And thus we see that the Cartan matrix $\mathcal{C}$ determines the entire structure of the Lie algebra $\mathcal{L}$ [Humphreys, 1972; Gilmore, 1974].

These structure equations hold for any simple Lie algebra, but the McKay correspondence holds only for the ADE Lie algebras.

Note that these structure equations hold also for the affine Lie algebras. In this case, the indices $\{i, j\}$ run from 1 through $r+1$, where r is the rank of the Coxeter graph. This implies that the Cartan matrix is based on the extended Coxeter graph (which has an extra node) [Hiller, 1982; Kac, 1985].

The extended Coxeter graph generates an infinite order Coxeter group, and this graph also corresponds to an infinite dimensional affine Lie algebra. These structures will be described in Chap. 5.

There is a second ADE correspondence theorem proved by McKay, which he called Proposition 2.

McKay Proposition 2. *Each ADE affine Coxeter graph, labeled with balance numbers, corresponds to a finite subgroup m of $\mathcal{SU}(2)$, such that m has a two-dimensional representation $R[2]$ (provided by the fundamental representation of $\mathcal{SU}(2)$) which obeys the tensor product formula:*

$$R[2] \otimes R[i] = \Sigma n[i,j] R[j] \text{ (with sum } \Sigma \text{ over } j),$$

where i is a balance number labeling a node of the affine Coxeter graph, so that $R[i]$ is an irrep of m and $n[i,j]$ is the number of links joining node i to node j [McKay, 1980; 1981].

For example, take the E_7 affine Coxeter graph:

```
1–2–3–4–3–2–1
      |
      2
```

For $R[2] \otimes R[4]$ we have: $n[4,3] = 2$ and $n[4,2] = 1$, all other linkages yielding 0. So the sum is $2(3) + 1(2) = 8$, which is the dimension of the tensor product space $R[2] \otimes R[4]$.

Note that this works out because every balance number B obeys the formula: $2B =$ the sum of the balance number linked to B.

And this formula is true for every ADE affine Coxeter graph.

However, the deeper reason for this curious symmetry is the McKay correspondence as expressed in his Proposition 2.

One consequence of the McKay correspondence is that the center of the group algebra of any McKay group is isomorphic to the Cartan subalgebra of the corresponding affine ADE Lie algebra.

In the E_7 case, this means that the 8D center of $\mathbb{C}[\mathcal{OD}]$ is isomorphic to the 8D Cartan subalgebra of the affine E_7 Lie algebra.

These claims need some unpacking. The center of any group or algebra is the set of elements that commute with all the elements of the group or algebra.

For a simple Lie algebra, the center is typically a discrete set of a small number of elements such as the identity element or $\{\pm 1\}$. More important to the overall structure of the Lie algebra is the Cartan subalgebra, the largest commutative subalgebra of the Lie algebra. In the case of the compact form of a simple Lie algebra, of rank n, the Cartan subalgebra is $\mathfrak{R}^n$, which generates the torus T^n as the Cartan subgroup of the corresponding compact Lie group.

For the affine form of the rank-n Lie algebra, the extended Coxeter graph implies that the Cartan subalgebra is $\mathfrak{R}^{n+1}$.

The Cartan subalgebra is a very rich structure with a Coxeter arrangement of hyperplane mirrors, whose roots generate a lattice that locates the eigenvalues of the generating operators.

This is the very heart of quantum theory, according to which only the eigenvalues of measurement operators can be actualized from the realm of the possible [Heisenberg, 1958].

With this motivation we will look more closely at the structure of Lie groups and Lie algebras in the next chapter.

Appendix A (Cf. Chap. 3)

The affine E_7 Coxeter graph, with balance numbers (including the extra number 1*), and with indexing numbers in italics is:

```
 0  1 2 3 5 6 7
1*–2–3–4–3–2–1
       |
       2
       4
```

Thus the affine E_7 Cartan matrix, $\mathcal{C} = 2\mathcal{E} - \mathcal{A}$, where $\mathcal{E}$ is the identity matrix and $\mathcal{A}$ is the adjacency matrix is:

	0	*1*	*2*	*3*	*4*	*5*	*6*	*7*
0	2	−1	0	0	0	0	0	0
1	−1	2	−1	0	0	0	0	0
2	0	−1	2	−1	0	0	0	0
3	0	0	−1	2	−1	−1	0	0
4	0	0	0	−1	2	0	0	0
5	0	0	0	−1	0	2	−1	0
6	0	0	0	0	0	−1	2	−1
7	0	0	0	0	0	0	−1	2

According to McKay, the eigenvectors of this Cartan matrix are the columns of the $\mathcal{OD}$ character table displayed below.

The character table for the OD group
(8 classes, $\mathcal{C}_i$, & 8 iireps, $\mathcal{R}_i$).

		$1\mathcal{C}_0$	$6\mathcal{C}_1$	$8\mathcal{C}_2$	$12\mathcal{C}_3$	$6\mathcal{C}_4$	$1\mathcal{C}_5$	$8\mathcal{C}_6$	$6\mathcal{C}_7$
	$\mathcal{R}_0$	1	1	1	1	1	1	1	1
→	$\mathcal{R}_1$	2	0	1	0	$\sqrt{2}$	−2	−1	$-\sqrt{2}$
	$\mathcal{R}_2$	3	−1	0	−1	1	3	0	1
→	$\mathcal{R}_3$	4	0	−1	0	0	−4	1	0
	$\mathcal{R}_4$	2	2	−1	0	0	2	−1	0
	$\mathcal{R}_5$	3	−1	0	1	−1	3	0	−1
→	$\mathcal{R}_6$	2	0	1	0	$-\sqrt{2}$	−2	−1	$\sqrt{2}$
	$\mathcal{R}_7$	1	1	1	−1	−1	1	1	−1

Note that the number of elements in each class is indicated along with the class label. These numbers add up to the order of the $\mathcal{OD}$ group: 48. This order is also the sum of the squares of the iirep dimensions: $1 + 4 + 9 + 16 + 4 + 9 + 4 + 1 = 48$. The arrows indicate the three representations beyond the five iireps of the octahedral group (of order 24).

Appendix B

McKay Groups as Quaternion Group Analogues
[Coxeter and Moser, 1963; Sirag, 2008]

Label:	Coxeter graph:	Coxeter $\langle p, q, r\rangle$ label	McKay subgroup of $\mathcal{SU}(2)$ (order) and quaternion formula
A_n	o–o–...–o	none	$\mathcal{Z}_{n+1}$ $(n+1)$ $\mathbf{a}^{r+1}=\mathbf{1}$
D_n	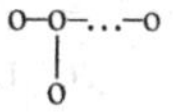 o–o–...–o | o $(n \geq 4)$	$\langle 2, 2, r-2\rangle$	$\mathcal{Q}_n$ $4(n-2)$ $\mathbf{a}^2 = \mathbf{b}^2 = \mathbf{c}^{r-2} = -\mathbf{1}$
E_6	o–o–o–o–o | o	$\langle 2, 3, 3\rangle$	$\mathcal{TD}$ (24) $\mathbf{a}^2 = \mathbf{b}^3 = \mathbf{c}^3 = -\mathbf{1}$
E_7	o–o–o–o–o–o | o	$\langle 2, 3, 4\rangle$	$\mathcal{OD}$ (48) $\mathbf{a}^2 = \mathbf{b}^3 = \mathbf{c}^4 = -\mathbf{1}$
E_8	o–o–o–o–o–o–o | o	$\langle 2, 3, 5\rangle$	$\mathcal{ID}$ (120) $\mathbf{a}^2 = \mathbf{b}^3 = \mathbf{c}^5 = -\mathbf{1}$

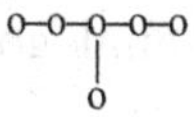

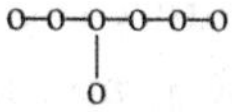

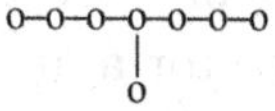

Here we view $\mathcal{SU}(2)$ as a 3-sphere the set of all unit-length quaterions. The quaternions constitute a 4D vector space, discovered by William Rowan Hamilton in the 19th century. Hamilton was attempting to find the mathematical formulas for describing the rotations in 3D space as an analog to using the complex numbers to describe rotations in 2D space. It took him 15 years to make his great discovery that he needed rotation formulas with four variables, which he called quaternions. The flash of insight came on October 16, 1843, while walking with his wife in a Dublin park. On a stone of Brougham bridge, he carved with his penknife the defining formula for quaternions:

$$\mathbf{i}^2 = \mathbf{j}^2 = \mathbf{k}^2 = \mathbf{ijk} = -\mathbf{1},$$

where $\mathbf{i}$, $\mathbf{j}$ and $\mathbf{k}$ are square-roots of minus one [Hankins, 1980].

Implicit in Hamilton's quaternion formula is the finite quaternion group, consisting of the eight elements $\{\mathbf{1}, -\mathbf{1}, \mathbf{i}, -\mathbf{i}, \mathbf{j}, -\mathbf{j}, \mathbf{k}, -\mathbf{k}\}$. In the McKay groups listed above, the quaternion group corresponds to $\mathcal{Q}_4$, in the sense that the D_4 graph corresponds to Hamilton's quaternion formula:

$$\text{D}_4 \text{ graph} \quad \longleftrightarrow \quad \mathbf{i}^2 = \mathbf{j}^2 = \mathbf{k}^2 = \mathbf{ijk} = -\mathbf{1}$$

The D_4 graph is an abstract image of the quaternion formula, in the sense that we can count nodes from the central node out to three legs of the graph, and get the set {2, 2, 2} of the powers in the quaternion formula. This matches Coxeter's label $\langle 2, 2, 2\rangle$ for $\mathcal{Q}_4$.

In similar fashion, all the D and E type McKay groups can be derived from the corresponding Coxeter graph. The elements labeled **a**, **b** and **c** are McKay group generators, which are analogous to the generating role of the quaternions {**i**, **j**, **k**}.

The A-type McKay group $\mathcal{Z}_{n+1}$ is a degenerate case, in which we just count the number of nodes in the A-type graph and write the group formula as $\mathbf{a}^{n+1} = \mathbf{1}$. All the elements of the cyclic group $\mathcal{Z}_{r+1}$ with powers {1 through $n+1$} of the generator **a** are elements of $\mathcal{Z}_{n+1}$.

Note also that the formula for Z_{r+1} has **1** rather than $-$**1**, and this implies that the A-type subgroups of $\mathcal{SO}(3)$ are the same as the A-type subgroups of $\mathcal{SU}(2)$, which double covers $\mathcal{SO}(3)$.

Moreover, if we rewrite the formulas for the D and E type McKay groups with **1** instead of $-$**1**, we will have the formulas for the corresponding subgroups of $\mathcal{SO}(3)$. In fact, Coxeter's labels for these groups simply replace $\langle p, q, r\rangle$ with (p, q, r). Thus:

$$\mathcal{SO}(3) = \mathcal{SU}(2)/\{\pm\} \longrightarrow (p, q, r) = \langle p, q, r\rangle/\{\pm 1\}.$$

For example $\langle 2, 2, 2\rangle$ is $\mathcal{Q}_4$, so that (2, 2, 2) is $\mathcal{K}_4$, the Klein-4 group. Also we have: $\langle 2, 3, 4\rangle/\{\pm 1\} = (2, 3, 4)$, so that $\mathcal{OD}/\{\pm 1\} = \mathcal{O}$.

Note that Coxeter's (p, q, r) is derived from the only set of solutions to the Diophantine equation:

$$1/p + 1/q + 1/r > 1$$

[Gilmore, 1974, p. 312]

Chapter 5

Lie Groups and Lie Algebras

In the context of unified field theory, the most important Lie groups and Lie algebras are those of ADE type. In fact, it is standard practice in the context of unified string theory to propose the series of projections [Green and Gross, 1986]:

$$\mathrm{E}_8 \otimes \mathrm{E}_8 \Rightarrow \mathrm{E}_6 \Rightarrow \mathrm{D}_5 \Rightarrow \mathrm{A}_4,$$

where A_4 is the label for the Lie algebra $\mathbf{su}(5)$, whose Lie group is $\mathcal{SU}(5)$, with the maximal subgroup $\mathcal{U}(1) \otimes \mathcal{SU}(2) \otimes \mathcal{SU}(3)$, which is recognized as the product of the gauge groups for the three forces: electromagnetism, weak force, and strong (color) force.

Indeed, $\mathcal{SU}(5)$ is the minimal group structure for a unification of these forces as proposed in 1974 by [Georgi and Glashow]. Moreover, in 1975 the unification gauge group $\mathcal{SO}(10)$, corresponding to D_5, was proposed by Fritzsch and Minkowski [Ross, 1984].

The series of Lie group projections to Lie subgroups is modeled by the removal of nodes from the ADE graphs (since the lower rank graphs are subgraphs of the higher rank graphs):

```
     E8                    E6             D5          A4

0-0-0-0-0-0-0   =>   0-0-0-0-0   =>   0-0-0-0   =>   0-0-0-0
    |                    |              |
    0                    0              0
```

The group $\mathrm{E}_8 \otimes \mathrm{E}_8$ was central to the development of heterotic string theory by the Princeton String Quartet [Gross, Harvey, Martinec, and Rohm, 1985]. This was a follow-up to the discovery that either $\mathrm{E}_8 \otimes \mathrm{E}_8$ or $\mathcal{SO}(32)$ (i.e. D_{16}) would cancel anomalies in 10D superstring theory, so that such a theory would be self consistent [Green and Schwarz, 1984].

This extraordinary result launched a tidal wave of further discoveries in superstring theory, which culminated in 1995 with the unification of the five competing string theories by way of various dualities within the overarching M-theory [Witten, 1995]. This is an 11D theory whose "low energy" limit is 11D supergravity as developed in the '70s [Nieuwenhuizen and Freedman, 1979]. This theory, which is a point particle theory rather than a string theory, had E_7 as gauge group [Howe, 1979], which could be seen as a projection between E_8 and E_6 in the above series (where E_7 is skipped over).

Motivated by the striking usefulness of these ADE Lie algebras, we will now look more closely at the structure of Lie groups and Lie algebras, and their intimate relationship to each other. Of course, this is an immense topic with many excellent texts, so that only a sketch describing the most salient structures are provided here [Gilmore, 1974; Humphreys, 1972].

A Lie group G is a smooth space which is also a group. Thus G is a differentiable manifold, each point of which is an element of a group, in the sense that the four basic axioms of group theory are obeyed by these elements. (Lie groups are named for Sophus Lie, 1842–1899.)

Closure: $\mathbf{a} \circ \mathbf{b} = \mathbf{c}$, where **a, b** and **c** are points of the group manifold. Also, the binary operation $\circ$ means that (on the group manifold) the point **a** transforms **b** into the point **c**.

Associativity: $\mathbf{c} \circ (\mathbf{a} \circ \mathbf{b}) = (\mathbf{c} \circ \mathbf{a}) \circ \mathbf{b}$

Identity: $\mathbf{e} \circ \mathbf{a} = \mathbf{a} = \mathbf{a} \circ \mathbf{e}$ (where **e** is the identity element)

Inverse: $\mathbf{a} \circ \mathbf{a}^{-1} = \mathbf{e} = \mathbf{a}^{-1} \circ \mathbf{a}$ (where $\mathbf{a}^{-1}$ is the inverse of **a**)

Commutativity: $\mathbf{a} \circ \mathbf{b} = \mathbf{b} \circ \mathbf{a}$

(only true for commutative groups, also called Abelian groups in honor of Niels Abel, 1802–1829).

All compact simple Lie groups (including the ADE groups) are non-commutative, with commutative subgroups, the largest of which is a hypertorus T^n (a product of n circles), where n is the rank of the simple group. This largest commutative subgroup is of great structural importance and is called the Cartan subgroup of the Lie group.

A Lie group, as a transformation group, also acts on vector spaces of various dimensions, which are spaces external to the Lie group. These transformations are realized as various sets of $d \times d$ matrices acting on d-dimensional vector spaces. Such realizations are called the various representations of the Lie group.

The two most important representations are the fundamental and the adjoint representation. The fundamental represention is the smallest faithful representation and is typically the representation used to define the group.

For example, the A_n type group has a compact form called $\mathcal{SU}(n+1)$, which is defined as the set of all $(n+1)$-dimensional special unitary matrices. This group acts by definition on the complex $(n+1)$-dimensional vector space; and it simply rotates, without stretching, the vectors in that space. As a compact differentiable manifold, $\mathcal{SU}(n+1)$ has dimensionality $(n+1)^2 - 1$. And thus A_1 corresponds to $\mathcal{SU}(2)$, which is a 3D manifold; A_2 corresponds to $\mathcal{SU}(3)$, an 8D manifold. Geometrically, $\mathcal{SU}(2)$ is the 3-sphere S^3; and $\mathcal{SU}(3)$ is the product manifold, $S^3 \otimes S^5$. Also $\mathcal{SU}(n+1)$, corresponding to A_n, has the geometrical structure of the product of n spheres, of increasing odd dimension. (See also the Appendix below.)

Moreover, any ADE Lie group of rank n has a compact manifold form which is a product of n spheres, whose dimensions can be calculated from the corresponding Coxeter group invariants [Hiller, 1982]. So this is another connection between the ADE Lie groups and the ADE classified Coxeter groups (cf. Chap. 6).

The adjoint representation of a simple Lie group employs matrices of dimensionality equal to that of the Lie group itself. As a transformation group, the adjoint representation acts on the Lie algebra of the Lie group. This is because we may picture the Lie algebra as the tangent space T_eG at the identity element e of the Lie group G.

A tangent space at any point of a manifold is the vector space of all the tangent vectors at that point; and this tangent vector space is therefore of dimensionality equal to that of the manifold itself. The algebraic properties of this tangent space are induced by the group properties of the Lie group manifold.

The automorphism of G into itself induces a group adjoint action:

$$Ad_g(X) = gXg^{-1}, \quad \text{for all } g \in G, \quad \text{and} \quad X \in T_eG.$$

The derivative (i.e. the tangent) of this mapping at the identity e of G is an endomorphism of the Lie algebra $\mathfrak{g}$, which implies the adjoint representation of the Lie algebra:

$$ad_x(y) = [x, y] = xy - yx, \quad \text{for all } x, y \in \mathcal{G} = T_eG.$$

The great importance of the adjoint representation of $\mathcal{G}$ is that it provides the structure constants of $\mathcal{G}$, which are defined via the Lie product on the basis vectors $\{E^i\}$ of $\mathcal{G}$:

$$[E^i, E^j] = \sum_k C_k^{ij} E^k$$

so that the matrix elements for $\mathrm{ad}_{E}i$ are defined by:

$$[ad_E i]_k^j = C_k^{ij}$$

where C^{ij} are elements of the Cartan matrix as in Chap. 4.

Note that Lie algebra elements (as in any matrix algebra) have both additive and multiplicative properties, as exemplified in the Lie product $[x, y] = xy - yx$.

There is an exponential mapping of the Lie algebra to the Lie group:

$$\exp(\mathcal{G}) = G$$

which can be realized via a power series on matrices.

$$\exp(x) = \Sigma \frac{(x)^n}{n!},$$

where the sum over n runs from 0 to ∞.

For example the set of all $n \times n$ unitary matrices (i.e. that in which any matrix times its complex-conjugate transpose is equal to 1) forms the $\mathcal{U}(n)$ group; and each unitary element u is generated by:

$$\mathcal{U} = \exp(i\mathcal{H}),$$

where $\mathcal{H}$ is an $n \times n$ Hermitian matrix (equal to its complex-conjugate transpose, $\mathcal{H} = \mathcal{H}^t$):

$$\mathcal{H}_{ij} = \mathcal{H}_{ji}^*.$$

Thus the Lie algebra $\mathbf{u}(n)$ of the Lie group $\mathcal{U}(n)$ is the set of $n \times n$ anti-Hermitian matrices, in which $\mathcal{H} = -\mathcal{H}$ [Schiff, 1968; Schutz, 1980].

The great importance of Hermitian matrices for both mathematics and physics lies in the fact that although these matrices have complex components, their eigenvalues are real numbers. And we note that in particle physics, the eigenvalues of certain ADE-type Hermitian matrices are the charges associated with eigenvectors, which are particle types.

Since there is a tangent space $T_g G$ at each element g of G, we can form the tangent bundle of G, which is the product manifold, of dimensionality 2 (dim. of G):

$$G \otimes \mathcal{g}.$$

This corresponds to the fact that $T_e G$ is equivalent to the set of left-invariant vector fields on the Lie group manifold G. This equivalence can be seen as a consequence of the fact that any vector in $T_e G$ can be parallel-transported along G by action of the elements of G (acting on the left) to all points of G, thus creating a left-invariant vector field. This produces an isomorphism between $T_e G$ and the set of all left-invariant vector fields on G. There is, of course, a similar isomorphism between $T_e G$ and the set of all right-invariant vector fields. (See the Appendix on Parallelizable Spheres.)

The tangent bundle is an example of a vector bundle, which is a fiber bundle whose fibers are vector spaces. A tangent bundle has a base space B, a base space B, each point x of which, has the tangent space $T_x B$ consisting of all tangent vectors at x. For a base space B of real dimension n, the structure group of the tangent bundle is $\mathcal{GL}(n, \mathfrak{R})$, the set of all general linear $n \times n$ matrices of real numbers. So for the base space as a Lie group G, where G is a compact Lie group of real dimension n, any fiber is a copy of the Lie algebra $\mathcal{G}$ and the structure group of the tangent bundle is $\mathcal{GL}(n, \mathfrak{R})$.

If G were a complex Lie group, the structure group would be $\mathcal{GL}(n, \mathbb{C})$. Such a group has the very important subgroup $\mathcal{SL}(n, \mathbb{C})$ of special linear $n \times n$ complex matrices (where "Special" means that these matrices have determinent 1). And we note that these groups are classified by the A-type Coxeter–Dynkin graphs, so that A_k corresponds to $n = k + 1$ in the Lie group $\mathcal{SL}(n, \mathbb{C})$, which is a non-compact group whose compact form is $\mathcal{SU}(n)$.

We note in passing that in general relativity (Einstein's theory of gravity), the tangent bundle on curved spacetime provides a connection between local coordinate frames. This connection, describing the parallel transport of tangent vectors, is an affine connection called the Levi–Civita connection and is formalized by the Christoffel symbol Γ^k_{ij} [Schutz, 1980; Wald, 1984]. Moreover, general relativity can be formalized as a classical gauge field theory, whose gauge group is $\mathcal{SL}(2,\mathbb{C})$, with Coxeter–Dynkin label A_1 [Carmeli, 1982].

However, in this book we will treat gravity in the context of the quantum gravity afforded by supersymmetry and string theory.

A fiber bundle can be considered as a generalization of a Lie group, in the sense that any Lie group G with a closed subgroup H can be viewed as a fiber bundle with G/H as its base space, and H as fiber, a copy of which is attached to each point of the base G/H, which is the space of cosets of H within G. In general, there will not be a "trivial" section S, where each point of S is a point of H, corresponding to a point of G/H. Such a section is called trivial since it can be globally defined.

In general, a fiber bundle is a manifold with the projection:

$$\pi \colon M \to B$$

(with standard fiber F and structure group G) such that the inverse image $\pi^{-1}(U)$ is $U \times F$ (where U is a member of a family of homeomorphisms of open sets in B). And G acts as group of automorphisms of F.

A Principal Fiber Bundle (PFB) has $F = G$. And associated with the PFB is a set of vector bundles in each of which F is some vector space acted on by some representation of G.

Note that this Lie group fiber-bundle structure is an analog of the octahedral group structure $\mathcal{O} \to \mathcal{O}/\mathcal{K}_4$ whose "fiber" $\mathcal{S}_3$ (the symmetric-3 group) consists of the six cosets of $\mathcal{K}_4$ within $\mathcal{O}$. (Cf. Chap. 2.) Since $\mathcal{O} = \mathcal{K}_4 \otimes \mathcal{S}_3$, we have here an analog of a fiber bundle with a trivial section.

A fiber-bundle with trivial section is especially straightforward in the case of a semi-simple Lie group, which has the structure of a product of simple Lie subgroups. In this case, any of these subgroups can be considered as a base space, and the other groups making up the product will function as the fiber in the bundle made up of the full set of subgroups in the product group.

In application to particle physics, the PFB is a trivial bundle $G \otimes B$, where G is the gauge group (i.e. fiber) and B is spacetime. G can also be a product of gauge groups, such as $\mathcal{U}(1) \otimes \mathcal{SU}(2) \otimes \mathcal{SU}(3)$, with the bundle structure remaining trivial.

Associated with the PFB is a set of vector bundles. In each such vector bundle, the base space is the same as that of the principal bundle, while each fiber is a copy of the vector space corresponding to some representation of the Lie group which is the fiber of the principal bundle. Thus the Lie group acts as a symmetry group on the vector spaces in the associated vector bundle. This symmetry group is called the structure group of the vector bundle [Bleecker, 1981].

In the case of the adjoint representation of the Lie group G, the vector bundle associated with G has the Lie algebra $\mathcal{G}$ as fiber.

In application to particle physics gauge theory, it is customary to use the PFB, in which the fiber is the gauge group. Since the gauge group is a Lie group, whose elements can be generated by the exponential mapping of the Lie algebra, we can write the action of the gauge group $\mathcal{U}$ on the particle wave function as:

$$\mathcal{U}\Psi = \exp[-iq\Sigma\theta^k(x)F_k]\Psi,$$

where the sum Σ is over k; x is an element of the spacetime base; q is the "coupling constant" for the gauge field, such as electric charge for the gauge group $\mathcal{U}(1)$. F_k is a basis element of the Lie algebra of the gauge group. The parameter θ^k is a continuous function of spacetime, and is thus called a local gauge field. A "local" transformation of the gauge group is distinguished from a "global" transformation, in which there is no change in θ^k for all points of spacetime. Mathematically the gauge field is called a connection on the fiber bundle structure [Bleecker, 1981].

The structure of a simple Lie group is determined by the structure of the Cartan subalgebra of the Lie algebra, which generates the Cartan subgroup of the Lie group. In the case of a compact Lie algebra, this Cartan subgroup is a hypertorus T^n equivalent to the product of n copies of $U(1)$, where n is the rank of the Lie algebra (i.e. the number of nodes in the Coxeter graph of the Lie algebra). This is true for all simple Lie algebras, but we are here thinking especially of the ADE Lie algebras (in which the Coxeter graph and the Dynkin diagram are identical).

The Cartan subalgebra of the Lie algebra $\mathcal{g}$ is a vector space $\mathfrak{h} = \mathfrak{R}^n$ (where n is the rank of $\mathcal{g}$). This is a very special vector space, whose basis vectors are operators $(\mathfrak{h}_1, \ldots, \mathfrak{h}_n)$ which act on all the basis elements of $\mathcal{g}$:

$$[\mathfrak{h}_i, \mathcal{E}_k] = \mathfrak{h}_i \mathcal{E}_k - \mathcal{E}_k \mathfrak{h}_i = \lambda_i \mathcal{E}_k,$$

where this equation is the Lie algebra form of the ordinary eigenvalue equation of an operator O acting on a vector v:

$$Ov = \lambda v.$$

In the Lie algebra case, $\mathcal{E}_k$ is a basis element of the non-commutative part of $\mathcal{g}$, so that λ_i is an eigenvalue of $\mathfrak{h}_i$, while $\mathcal{E}_k$ is the eigenvector associated with λ_i.

Also as the Cartan subalgebra basis elements are commutative, we have:

$$\mathfrak{h}_i \mathfrak{h}_k - \mathfrak{h}_k \mathfrak{h}_i = 0$$

so that 0 is the eigenvalue of $\mathfrak{h}_i$ for the eigenvector $\mathfrak{h}_k$, and vice versa.

As a simple example, take the case of A_2. This is the Lie algebra $\mathfrak{su}(3)$, which is an 8D algebra with a 2D Cartan subalgebra $\mathfrak{h}$. For $\mathfrak{h}_1$ and $\mathfrak{h}_2$ as basis elements of $\mathfrak{h}$, and $\mathcal{E}_1$ as one of the six basis elements of the non-commutative portion of $\mathfrak{su}(3)$, we have:

$$\begin{aligned} [\mathfrak{h}_1, \mathfrak{h}_2] &= [\mathfrak{h}_2, \mathfrak{h}_1] = 0, \\ [\mathfrak{h}_1, \mathcal{E}_1] &= 1\mathcal{E}_1, \\ [\mathfrak{h}_2, \mathcal{E}_1] &= 0\mathcal{E}_1. \end{aligned}$$

Thus $\mathfrak{h}_1$ and $\mathfrak{h}_2$ are commutative and are eigenvectors of each other with eigenvalue 0. And $\mathcal{E}_1$ is an eigenvector of both $\mathfrak{h}_1$ and $\mathfrak{h}_2$ with eigenvalues 1 and 0, respectively. The root corresponding to $\mathcal{E}_1$ is a vector in the space $\mathfrak{h}^*$ (dual to $\mathfrak{h}$), with coordinates 1 and 0.

Let α be the root with coordinates 1 and 0. Then α will be one point in the hexagonal root structure (Cf. Chap. 4).

The six roots, along with their eigenvalues, are [Georgi, 1982, p. 63]:

$$\begin{array}{ll} \alpha(1,0); & -\alpha(-1,0); \\ \beta(-1/2,\sqrt{3/2}); & -\beta(1/2,-\sqrt{3/2}); \\ \alpha+\beta(1/2,\sqrt{3/2}); & -\alpha-\beta(-1/2,-\sqrt{3/2}). \end{array}$$

There are two basic roots, α and β, corresponding to the A_2 Coxeter graph o–o, in which the link indicates a 120° angle between these two basic roots. The A_2 graph also indicates a 60° angle between the two mirror planes, which in this case are straight lines orthogonal to each basic root. By mirror reflection the two basic roots generate four more roots ($-\alpha, -\beta, \alpha+\beta$, and $-\alpha-\beta$). Also the two basic mirrors generate a third mirror.

This structure is exactly like that of the toy Kaleidoscope with five reflection chambers generated from the fundamental chamber, so that a 6-sided snowflake design is obtained.

Note that it is the 2D viewing screen of the physical Kaleidoscope that corresponds to the A_2 root and mirror structure.

Mathematically, the roots correspond to the adjoint representation of the Lie algebra. There is a root lattice generated by integer sums of the basic roots. And this root lattice is a sublattice of the weight lattice whose points are called weights. These weights carry the eigenvalues of other representations of the Lie algebra.

Most importantly, in the Standard Model of particle physics, the roots correspond to gauge particle (bosonic) eigenvalues, while the weights correspond to matter particle (fermionic) eigenvalues. Ordinarily, Pauli's "spin-statistics theorem" [Pauli, 1940; Streater and Wightman, 1964] governs the contrast between the bosonic (integral spin) statistics and the fermionic (half-integral spin) statistics. However, at the fundamental particle level of the Standard Model this contrast is governed by the basic distinction between the root lattice and the weight lattice.

Moreover, the Cartan matrix can be viewed as a transformation between the root lattice and the weight lattice. For example, in the

A_2 case, the Cartan matrix is:

$$\begin{pmatrix} 2 & -1 \\ -1 & 2 \end{pmatrix}$$

so that the two basic roots (α_1, α_2) as vectors in terms of weight bases (λ_1, λ_2) are:

$$\alpha_1 = 2\lambda_1 - \lambda_2; \quad \alpha_2 = -\lambda_1 + 2\lambda_2.$$

If we invert the A_2 Cartan matrix, we have:

$$(1/3)\begin{pmatrix} 2 & 1 \\ 1 & 2 \end{pmatrix}$$

so that: $\lambda_1 = (1/3)(2\alpha_1 + \alpha_2); \lambda_2 = (1/3)(\alpha_1 + 2\alpha_2)$, cf. [Humphreys, 1972, p. 68].

Continuing with the A_2 case, the fundamental representation of $\mathfrak{su}(3)$, is three-dimensional. There are two basic weights that generate a weight lattice, which embeds the root lattice (cf. Chap. 4).

The eigenvalues of the 3D representation are carried by three weights forming a triangle, and also (by reflection) an anti-triangle. These two triangles form a "Star of David" pattern in the 2D space $\mathfrak{h}^*$ dual to the 2D Cartan subalgebra $\mathfrak{h}$.

In Gell-Mann's quark model, the Lie group $\mathcal{SU}(3)$ acts on the 3D vector space with three quark flavors as basis: up, down, and strange (u, d, s). In this theory a proton (uud) is made up of two up quarks and one down quark, while a neutron (udd) is made up of two down quarks and one up quark. The transformations of protons into neutrons (and vice versa) are a basic aspect of the strong nuclear reaction which keeps an atomic nucleus intact against the tendency of the nucleus to break apart due to the repulsion of the positive electric charges of the protons (like charges repel; unlike charges attract).

In the pre-quark nuclear model, protons and neutrons change into each other by exchanging π mesons. In the more precise quark model, a π meson consists of a quark and an antiquark (u, d*) (cf. Chap. 2).

According to Gell-Mann's conventions, the 2D eigenvalue space $\mathfrak{h}^*$ has as orthogonal basis vectors Isospin I_z and hypercharge Y. The weight diagram for the three quark (u, d, s) and three anti-quark (u*, d*, s*) states form two intersecting triangles (where I_z is horizontal and Y is vertical) [Kokkedee, 1969; Georgi, 1982].

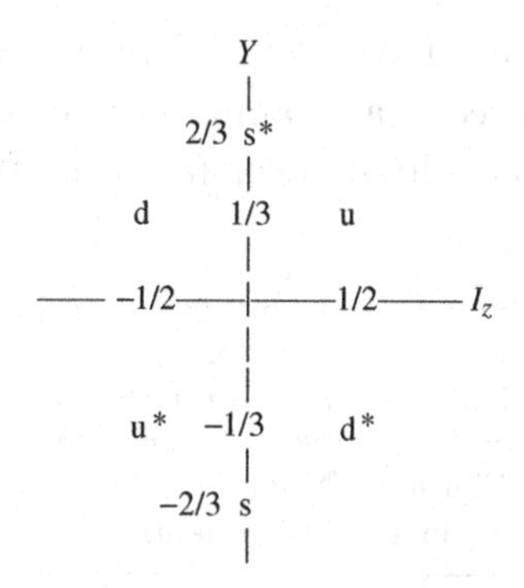

The lattice for this "Star of David" pattern is the weight lattice depicted in Chap. 4.

One especially striking aspect of the quark model is the assignment of fractional electric charges to the quarks: u (2/3) and d (−1/3).

Electric charge Q for these quarks is derived from the equation:

$$Q = I_z + Y/2$$

so that we have the electric charge assignments:

$$\begin{array}{llll} \text{u:} & 2/3 & \text{d:} -1/3 & \text{s:} -1/3 \\ \text{u*:} & -2/3 & \text{d*:} \ \ 1/3 & \text{s*:} \ \ 1/3. \end{array}$$

The quark model uses weight diagrams to assign combinations of u, d and s quarks to various representations of $\mathcal{SU}(3)$. Most notably, these are the diagrams for 8D and 10D representations of $\mathcal{SU}(3)$. The 10D representation was used to predict the existence of the Omega baryon with strangeness 3, that is, the state [s,s,s⟩. The discovery of this state in 1964 at Brookhaven, confirmed the great utility of the quark model [Georgi, 1982].

The later extension of the quark model to include the new quark **flavors:** charm, top and bottom, greatly complicated the picture.

Indeed the use of $\mathcal{SU}(3)$ as a quark flavor symmetry is now considered an "accidental symmetry" due to the low masses of the three lightest quarks (u, d, s) compared with the scale of the strong interaction [Gross, 2004].

Moreover, the extension of the first lepton family to three families in parallel with the three quark families demanded an answer to the question: Why are there three flavor families of quarks and leptons?

Up	Electron
Down	Electron neutrino
Charm	Muon
Strange	Muon neutrino
Top	Tau
Bottom	Tau neutrino

This is still considered the deepest unsolved problem in the Standard Model of particle physics, which makes these states the fundamental matter-particle states. My proposed solution is the structure of the octahedral group, in which the cosets of the Klein-4 subgroup in the context of the five permutation classes (of orders 1, 3, 8, 6, 6) provide the structure of three cosets of even permutation elements (corresponding to bosonic particles) and three cosets of odd permutation elements (corresponding to fermionic particles) with two quarks and two leptons in each of three families (cf. Chap. 2).

However, in the original quark model there was another glaring problem. As quarks are fermions of spin 1/2, they must obey Fermi–Dirac statistical rules, which say that no two (or more) particles of the same type can be in the same quantum state. Yet the quark model assigns the state $|\mathrm{uud}\rangle$ to the proton, $|\mathrm{udd}\rangle$ to the neutron, and the triply strange state $|\mathrm{sss}\rangle$ to the Omega baryon.

The solution to this problem was to introduce the new quantum number, called color, so that each quark would carry one of three colors, conventionally called red, green and blue (r, g, b). This introduced the necessity of applying the $\mathcal{SU}(3)$ group as a gauge group for the strong color force. Since gauge force particles are bosons (spin 1) modeled by the adjoint representation, there must be eight of these $\mathcal{SU}(3)$ gauge particles, which are called gluons since they, in effect, glue the quarks together. Thus these eight gauge particles held quarks together as baryons such as neutrons and protons made up of three quarks. Gluons (carrying color and anti-color charges bind the quarks making up mesons. An example is a π meson made up of a quark and anti-quark (u, d*). It is assumed that these eight gluons are massless, which has been verified experimentally. This is analogous to the massless photon, which is the gauge particle for electromagnetism with $\mathcal{U}(1)$ symmetry.

Notice that we have shifted from $\mathcal{SU}(3)$ as the symmetry group acting on three flavors of quarks (u, d, s) and three anti-flavors (u*, d*, s*) to $\mathcal{SU}(3)$ as the gauge symmetry group acting on three colors of quarks (r, g, b) and three anti-colors (r*, g*, b*). This shift from flavor to color is accomplished by relabeling the states and also considering $\mathcal{SU}(3)$ as a gauge symmetry group.

This $\mathcal{SU}(3)$ theory of the strong color force is called Quantum Chromodynamics (QCD) by analogy with the very successful Quantum Electrodynamics (QED), which employs $\mathcal{U}(1)$ as its gauge group.

In 1973 QCD became a viable theory of the strong color force when Gross, Wilzek, and Politzer showed that the short range of the color force is due (very counter-intuitively) to the "asymptotic freedom" of this force. This means that the attraction between quarks strengthens with increasing distance and weakens with decreasing distance. Quite significantly, this strange behavior is largely due to the non-commutativity of the gauge group $\mathcal{SU}(3)$ [Gross, 1985]. This is in stark contrast with the long range electromagnetic force, which as a gauge force is due to the commutativity of the $\mathcal{U}(1)$ group.

The weak force gauge group $\mathcal{SU}(2)$ is also non-commutative, but the main reason for the short range of the weak force is the mass of the gauge particles (W^+, W^-, Z^0) which is due to the symmetry breaking of the Higgs mechanism, which has been confirmed by the discovery at the Large Hadron Collider of the Higgs particle in July 2012 [Carroll, 2012].

It is clear that $\mathfrak{h}^*$ (dual to the Cartan subalgebra $\mathfrak{h}$) is the space where all the particle symmetry diagrams of eigenvalues corresponding to eigenstates reside. The points of these diagrams locate vectors called roots and weights. In the context of particle physics, the roots carry bosonic (integral spin) eigenvalues. Thus the adjoint representation corresponds to gauge particles (spin-1). The weights carrry fermionic (half-integral spin) eigenvalues. Thus these are the eigenvalues for matter particles.

These eigenvalues are the charges carried by the various particles, both bosonic and fermionic. Thus $\mathfrak{h}^*$ the space in which the roots and weights live could be called charge space. It is ordinarily considered an "internal" space of the particles in order to distinguish it from spacetime, the "external" space of the particles.

Note also that the symmetric structure of the charge space diagrams is due to the action of the finite reflection group on the mirror hyperplanes orthogonal to the roots. So this charge space $\mathfrak{h}^*$ can also be called reflection space.

The number of roots is equal to the number of non-commutative basis elements of the Lie algebra $\mathfrak{g}$. This is because these roots carry the eigenvalues of the non-commutative basis elements. (There are also "zero roots" carrying the zero eigenvalues of the commutative basis elements of $\mathfrak{h}$.)

For an n-dimensional Cartan subalgebra $\mathfrak{h}$, there are n basic roots, which generate all the other roots by the action of the finite reflection group (called the Coxeter group, or Weyl group) of $\mathfrak{g}$. Thus there are n basic reflection hyperplanes (each of dimension $n-1$) called basic mirrors. Each mirror has two roots attached to it, one of which is the negative of the other, and the reflection group generates extra mirrors as well as the extra roots attached to them.

The set of roots is called a root system. And the set of mirrors is called a Coxeter arrangement. There are many applications of these structures, which will be described in later chapters.

The number X of roots, being in 1–1 correspondence with the non-commutative basis elements of $\mathfrak{g}$, can be calculated from the formula:

$$X = nK,$$

where n is the rank of $\mathfrak{g}$, and K is the Coxeter number of $\mathfrak{g}$. Thus, since n is the dimension of $\mathfrak{h}$, the total dimension of $\mathfrak{g}$ is:

$$n + nK.$$

Moreover, K is equivalent to the sum of the balance numbers on the extended Coxeter graph. For example, in the case of E_7, the extended Coxeter graph is:

$$\begin{array}{c} 1\text{–}2\text{–}3\text{–}4\text{–}3\text{–}2\text{–}1 \\ | \\ 2 \end{array}$$

Thus K is 18, and the dimension of $\mathcal{E}_7$ is:

$$7 + 7(18) = 133.$$

Since 7(18) is 126, there are 126 roots in the seven-dimensional Cartan subalgebra dual space $\mathfrak{h}^*$.

Although the extended Coxeter graph provides a way to calculate the dimension of the corresponding Lie algebra, the extended graph also relates to the structure of the affine Lie algebra. This is the simplest form of the more general family of Kac–Moody Lie algebras, which are infinite dimensional Lie algebras [Kac, 1985].

For each ADE Coxeter graph there is an extended graph, and thus an ADE affine Lie algebra.

Note also that for each ADE Coxeter graph, the balance number on the extended graph provide two more finite group orders:

(1) The McKay group order, which is the sum of the squares of the balance numbers. (For E_7, the MacKay group is the Octahedral Double group of order 48.)

(2) The product of the balance numbers is a factor in the order of the reflection group (i.e. Coxeter group). For E_7, the balance number product is 288, so that the E_7 Coxeter group has order:

$$288 \times 7! \times 2,$$

where 7! is the order of the symmetric-7 group which permutes the seven basic mirrors; and two is the number of 1's in the extended Coxeter graph, and is the order of the symmetric-2 group which expresses the bilateral symmetry of the extended E_7 graph itself.

For each ADE Coxeter graph there is an extended graph, and thus an ADE affine Lie algebra. The extended graphs were first described by Coxeter in classifying what he called affine reflection groups. These groups are of infinite order and have the finite reflection groups as largest finite subgroup, as will be described in more detail in the next chapter, cf. [Coxeter, 1973].

Appendix on Parallelizable Spheres

The Lie groups $\mathcal{U}(1)$ and $\mathcal{SU}(2)$ are special in the sense that as manifolds S^1 and S^3 they are the only spheres which are Lie groups. As these spheres are Lie groups they are parallelizable by way of their Lie algebra structures which are made up of all left-invariant vector fields on the Lie group manifold. It is a famous theorem of topology that S^2 is not parallelizable: it is impossible to comb the hair on a 2-sphere without making a cowlick. In fact, no even-dimensional sphere has a nowhere-zero vector field [Porteous, 1981].

Moreover, S^1 can be thought of as a one-dimensional torus T^1, so it is not surprising that the n-dimensional torus T^n (the product of n copies of S^1) is parallelizable. Indeed any Lie group manifold is parallelizable via the structure of the Lie algebra as the set of left-invariant vector fields on the group manifold.

Surprisingly, although no sphere beyond S^3 is a Lie group, the 7D sphere S^7 is parallelizable. This is due to the algebraic structure of the Octonions (also called Cayley numbers). It is a significant fact that the three parallelizable spheres correspond to the unit length elements of the three normed division algebras: Complex numbers, quaternions, and octonions:

$$\mathbb{C} \to S^1, \quad \mathfrak{H} \to S^3, \quad \mathbb{O} \to S^7.$$

While $\mathbb{C}$ is commutative, and $\mathfrak{H}$ is non-commutative, it is the lack of associativity in the octonions that keeps S^7 from being a Lie group.

It is quite interesting that the E_6, E_7 and E_8 Lie groups correspond to symmetries of the octonions [Dixon, 1994].

Chapter 6

Coxeter's Reflection Groups

Coxeter's reflection groups are the basis of all the ADE-classified mathematical objects. The ADE reflection groups are the most useful of all crystallographic reflection groups, which in turn are a subset of all reflection groups. In the 1930s Coxeter was able to classify all finite groups generated by reflections in Euclidean space of any number of dimensions [Coxeter, 1934, 1973; Hiller, 1982; Grove and Benson, 1985].

He accomplished this classification by devising a set of very simple graphs, whose nodes represented basic mirror hyperplanes, and whose links represented dihedral angles between these mirrors. Coxeter found that, these dihedral angles could only be π/p radians, where p is 3, 4, 5 or 6, corresponding to degrees of 60, 45, 36 or 30 respectively [Coxeter, 1973, p. 191].

The crystallographic reflection groups required the omission of $p = 5$. This corresponds to the pentagon in 2D space. Moreover, 2D is special in the sense that each n-sided polygon corresponds to a finite reflection group. However, among these polygon cases only triangles, squares and hexagons can be used to generate lattices, which are integer-sums of basis elements. This is consonant with the fact that for any dimension the ability to generate an integer-sum lattice is the basic condition for crystallographic reflection groups. In this sense, hyperdimensional crystals reside in hyperdimensional lattices.

Of seminal importance is the correspondence between crystallographic reflection groups and the classification of simple Lie algebras. In this context, these Coxeter reflection groups are called Weyl groups (cf. Chap. 4). These Lie algebra lattices (root lattices and weight lattices) live in the dual space to the Cartan subalgebra of dimension equal to the rank of the Dynkin graph (exactly like the Coxeter graph in the ADE cases).

The full set of crystallographic Coxeter graphs is depicted in the following two tables (where links with $\pi/3 = 60°$ are unmarked).

Crystallographic Coxeter graphs of ADE type.

A_n o, o–o, o–o–o ... D_n $n \geq 4$: o–o–o, o–o–o–o, o–o–o–o–o ...
(in each D_n graph an extra node o hangs below the second node)

E_6 o–o–o–o–o E_7 o–o–o–o–o–o E_8 o–o–o–o–o–o–o
(in each E graph an extra node o hangs below the third node)

Crystallographic Coxeter graphs of non-ADE type.

$B_n \geq 2$: o–4–o, o–4–o–o, o–4–o–o–o, o–4–o–o–o–o, ...

F_4 o–o–4–o–o

G_2 o–6–o

There are also Coxeter graphs corresponding to non-crystallographic Coxeter (reflection) groups. These are:

H_3 o–5–o–o H_4 o–5–o–o–o

$I_2(p)$: o–5–o, o–7–o, o–8–o, ... $(p > 8)$

Omitted here are the crystallographic groups that have previously been listed: $I_2(3) = A_2, I_2(4) = B_2$. $I_2(6) = G_2$, which correspond to triangular, square, and hexagonal lattices, respectively.

Note that the subset of ADE Lie groups corresponds to graphs which are called "singly laced." This is because Dynkin used multiple links instead of numbers to label the links [Humphreys, 1972]. Thus Coxeter's ADE graphs are identical to Dynkin's ADE graphs.

Moreover, Dynkin's classification entails C_n-type Lie groups, which, however, have the same Weyl-groups as the B_n-type Lie groups. The crystallographic Coxeter groups are identical to the Weyl groups, but the non-crystallographic Coxeter groups go beyond the Weyl group classification set.

Note that since Dynkin's Lie group diagrams and Coxeter's reflection group graphs were developed independently, the naturalness of these graph schemes seems evident.

Since in this book we are dealing with the Coxeter–Dynkin graphs of ADE type, we do not have to consider the non-ADE type graphs. However, it is possible to include G_2 and F_4 graphs, because G_2 can be derived by folding the affine E_6 graph, while F_4 can be derived by folding the affine E_7 graph [Hiller, 1982, pp. 35–36]. Indeed these groups are of some use in recent string theory developments. In the affine E_7 case, folding provides:

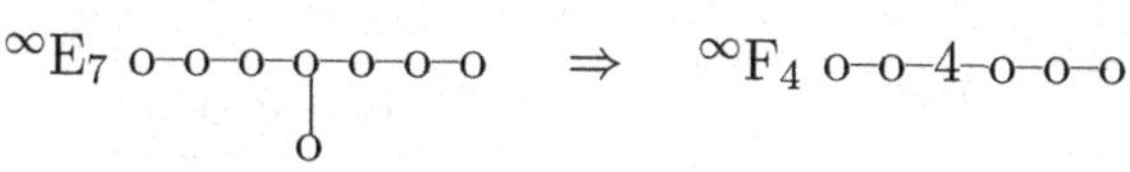

```
∞E7 o–o–o–o–o–o–o     ⇒     ∞F4 o–o–4–o–o–o
          |
          o
```

(by folding the left arm of the $^{\infty}E_7$ graph over the right arm). Cf. Appendix B: Folding ADE graphs.

Note that we have introduced the infinity symbol ∞ to indicate both the infinite order Coxeter group and the corresponding infinite dimensional affine Lie algebra.

Significantly, all the affine Coxeter graphs can be derived from the affine ADE graphs by folding the ADE graphs. These affine Coxeter graphs correspond to infinite discrete Coxeter groups, and also to infinite dimensional Lie algebras (the affine Lie algebras, which are the simplest of the Kac–Moody Lie algebras). The finite Coxeter groups and finite-dimensional Lie algebras correspond to sub-graphs of the affine graphs. Thus all the above graphs can be derived from the affine ADE graphs. In this sense, the ADE graphs unify the entire Coxeter graph structure.

It is also significant that the affine graphs are used in the proofs that the classification set of simple Lie algebras {A through G} is complete. These proofs depend on the derivation of the ordinary unextended graphs from the affine (extended) graphs [Gilmore, 1974; Georgi, 1982].

Here we present the complete set of affine ADE graphs (with balance numbers on the nodes) yielding finite ADE graphs corresponding to a finite ADE Coxeter groups:

```
∞An: 1–1, 1–1–1, 1–1–1–1,...      ⇒ An: o–o, o–o–o, o–o–o–o, ...
      \/   \ / \     /
      1     1       1

        1         1
        |         |
∞Dn: 1– 2–1, 1–2–2–1, ...         ⇒ Dn: o–o–o, o–o–o–o, ...
        |      |                          |      |
        1      1                          o      o

∞E6: 1–2–3–2–1                    ⇒ E6: o–o–o–o–o
         |                                  |
         2                                  o
         |
         1

∞E7: 1–2–3–4–3–2–1                ⇒ E7: o–o–o–o–o–o
         |                                  |
         2                                  o

∞E8: 2–4–6–5–4–3–2–1              ⇒ E8: o–o–o–o–o–o–o
         |                                  |
         3                                  o
```

Note that in the case of the ADE graphs, by summing the balance numbers on the affine graphs, we can generate the Coxeter number K, from which we can derive the dimension of the rank-n finite Lie algebra as $n+Kn$ (where n is the dimension of the Cartan subalgebra, and Kn is the dimension of the non-commutative portion of the Lie algebra).

$$\mathrm{A}_n\text{: } 2n+n^2 \quad \mathrm{D}_n\text{: } 2n^2-n \quad \mathrm{E}_6\text{: } 78 \quad \mathrm{E}_7\text{:}133 \quad \mathrm{E}_8\text{: } 248$$

We can also start from the finite Coxeter graphs and generate the affine graphs as extended Coxeter graphs. This harkens back to the Chap. 4 view of the balance numbers on the extended graphs as mirror vector magnitudes necessary to sum these vectors to zero, analogous to force vectors in balance [Hiller, 1982, p. 35]. In this picture, the extra node in the extended graph corresponds to a unit length mirror vector which balances all the other basic mirror vectors, that have been appropriately lengthened. The number of these basic mirror vectors is the rank of the graph, and these basic mirrors generate the finite Coxeter reflection group. The addition of an extra balancing vector corresponds to a mirror hyperplane, creating a closed mirrored chamber which generates an infinite-order reflection group.

One way to describe this infinite-order reflection group (also called the affine Weyl group) is to consider the lattice generated by the integer sum of the basic mirror vectors (called roots). The lattice points are elements in the infinite reflection group formed as the semi-direct product of the finite reflection group W with the lattice $L : W \varsigma L$ [Hiller, 1982, p. 192].

In the A_2 case, a toy Kaleidoscope with three mirrors forming an equilateral triangle in the viewing screen will demonstrate the infinity of reflections in the 2D plane. Removing one of these mirrors corresponds to the finite reflection (S_3 group) case, as a subgroup of the infinite reflection group, $S_3 \otimes L(\mathrm{A}_2)$. The six elements of the S_3 group provide the hexagonal "snowflake" designs in the ordinary (2-mirror) toy Kaleidoscope (cf. the A_2 lattice in Chap. 4). Coxeter emphasizes this analogy by calling the sets of hyperplane mirrors Kaleidoscopes. See "The Kaleidoscope" in [Coxeter, 1973].

In Chap. 4, the balance numbers on the extended graph is interpreted in the context of the McKay correspondence. Thus these numbers correspond to the iirep dimensions of the McKay group. For any ADE extended graph, these dimension numbers can be used to generate three fundamental structural numbers:

1. The sum of the iirep dimensions (the balance numbers) is the Coxeter number K, from which we can generate the Lie algebra dimension $D = n + nK$ (where n is the rank of the Coxeter graph).

2. The sum of the squares of the iirep dimensions is the order of the McKay group and thus the dimension of the McKay group algebra.

3. The product P of the iirep dimensions is a factor in the order of the Coxeter group: $P(n!)S$ where n is the rank of the Coxeter group, so that $n!$ is the order of the symmetry group acting on the basic mirror hyperplanes; and S is the order of the symmetry group acting on the extended Coxeter graph itself (which is equivalent to the number of 1's among the balance numbers). S is also the order of the center of the Coxeter group.

For example in the E_7 case, the extended Coxeter graph is:

```
1–2–3–4–3–2–1
      |
      2
```

Thus the E_7 Coxeter number is 18; the E_7 McKay group order is 48; and the E_7 Coxeter group order is $288(7!)(2) = 2{,}903{,}040$.

There is another way to calculate the order of the Coxeter group. For this we need to know the degrees of the homogeneous polynomial invariants of the Coxeter group. For a Coxeter group X of rank n, there are n fundamental invariants; and the homogeneous degrees of these invariants are called the fundamental degrees. The order of X is the product of these fundamental degrees. For example:

$$|E_7| = 2 \times 6 \times 8 \times 10 \times 12 \times 14 \times 18 = 2{,}903{,}040.$$

Note that the highest degree is 18, which is the E_7 Coxeter number.

The following table provides data for all the finite reflection groups.

Coxeter group table

Group	Order	Roots	Fundamental degrees
A_n	$(n+1)!$	n^2+n	$2,3,4,\ldots,n+1$
B_n	$(2^n)n!$	$2n^2$	$2,4,6,\ldots,2n$
D_n	$(2^{n-1})n!$	$2n(n-1)$	$2,4,6,\ldots,2(n-1),n$
E_6	51,840	72	2,5,6,8,9,12
E_7	2,903,040	126	2,6,8,10,12,14,18
E_8	696,729,600	240	2,8,12,14,18,20,24,30
F_4	1152	48	2,6,8,12
G_2	12	12	2,6
H_3	120	30	2,6,10
H_4	14,400	120	2,12,20,30
$I_2(m)$	$2m$	$2m$	$2m$

[Coxeter and Moser, 1965, p. 141; Grove and Benson, 1985, p. 76; Hiller, 1982, p. 87]

Note: This is the complete list of finite groups generated by reflections, as defined by Coxeter. The groups labeled A through G are crystallographic groups and thus correspond to Lie algebras with the same labels. In this Lie algebra context, these Coxeter groups are usually called Weyl groups. The C type Lie algebras happen to have the same Weyl group as the B type Lie algebras; and thus C does not appear in the list of Coxeter groups.

It must be noted that the labels A_n through $I_2(m)$ are modified Coxeter group labels [Hiller, 1982], in accordance with the fact that the crystallographic Coxeter groups correspond to Lie groups. Originally, Coxeter used labels which were more descriptive of the structure of the reflection groups and the corresponding Coxeter graphs.

Coxeter defined a reflection group of rank n as generated by n reflections R_i, where i runs from 1 through n. The Coxeter labels could be interpreted as corresponding to presentations of the reflection groups, as in the following examples [Coxeter and Moser, 1965]:

$$[\mathbf{3}^{n-1}]\text{: } (R_1)^2 = (R_2)^2 = \cdots = (R_{n-1})^2 = E \text{ (as identity element)}$$

with the following permutation cycles generating the Symmetric group of all permutations of $n+1$ things, S^{n+1}, whose order is

$(n+1)!$:

$$R_1 = (12), R_2 = (23), \ldots, R_{n-1} = (n-1n).$$

The Coxeter graph is the very simplest one (the unmarked link being understood as indicating an angle of $\pi/3$ between two mirrors):

$$\mathbf{o} - \mathbf{o} - \cdots - \mathbf{o} \text{ (with } n \text{ nodes and Lie group label } \mathrm{A}_n)$$

corresponding to the compact simple Lie group $\mathcal{SU}(n+1)$.

An important example is the group **[3,3]** corresponding to A_3. Therefore **[3,3]** is Coxeter's label for the symmetric-4 group, S_4, the symmetry group of the cube, and its dual, the octahedron. It is thus also called the octahedral group $\mathcal{O}$ (as described in Chap. 2).

[3,3,3] corresponds to A_4 and is thus the Weyl group of $\mathcal{SU}(5)$.

A more complicated example is:

$$\mathbf{[3,4,3]}\text{:}\ (R_i)^2 = (R_1R_2)^3 = (R_2R_3)^4 = (R_3R_4)^3 = E$$

(where E is the identity element in the presentation), and the Coxeter graph is:

o–o–4–o–o (corresponding to the F_4 Lie group label)

and where the four on the graph means an angle of $\pi/4$ between two mirrors; and it is understood that the unmarked graphs correspond to an angle of $\pi/3$ between two mirrors.

For graphs with three legs, more elaborate labels are necessary. See, for example, the E_7 Coxeter graph symbolism:

$$\begin{aligned}\mathbf{[3^{3,2,1}]}\text{:}\ (R_i)^2 &= (R_1R_2)^3 = (R_2R_3)^3 = (R_4R_5)^3 \\ &= (R_5R_6)^3 = (R_3R_7)^3\end{aligned}$$

which implies the indexing of the E_7 graph nodes as:

```
1   2   3   4   5   6
o---o---o---o---o---o
        |
        o   7
```

(where the "3, 2, 1" in the Coxeter label $[3^{3,2,1}]$ refers to the number of links in the three legs of the graph. The 3 at the beginning of the label refers to the $\pi/3$ angle between the mirrors indicated by the nodes.

The correspondence between the crystallographic Coxeter groups and the Lie groups is exemplified further by the intimate relationship between the fundamental degrees of the invariants of the Coxeter groups and the dimensionality of the Lie group manifold. This relationship is realized by the compact form of any n-rank simple Lie group, which has a manifold made up of the product of n spheres, whose dimension can be derived from the fundamental degrees according to the simple formula (where f is a fundamental degree):

$$2f - 1 = \text{ the dimension of any of the spheres.}$$

For example, in the case of E_7 (with Coxeter number 18) the fundamental degrees are:

$$2, 6, 8, 10, 12, 14, 18$$

so that the dimensionality of the E_7 group manifold (as a direct product of spheres) is:

$$3 + 11 + 15 + 19 + 23 + 27 + 35 = 133.$$

Note that 133 can also be derived from $7+18(7) = 7+126$, where 7 is the rank (and thus the dimensionality of the Cartan subgroup) and 126 is the number of *roots* (attached front and back to each of the 63 mirrors).

These 126 roots, while residing in the 7D dual space to the E_7 Cartan subalgebra, correspond to the dimensionality of the noncommutative part of the E_7 Lie algebra.

It is significant that each set of fundamental degrees (for the entire list of finite reflection groups) begins with the degree 2. In the case of the Lie group manifolds (corresponding to the crystallographic Coxeter groups, A_n through G_2), this implies that each compact Lie group manifold, made up of a direct product of spheres, has S^3 as a submanifold. Moreover, S^3, as the $\mathcal{SU}(2)$ manifold of the A_1 Lie group, is the only sphere (other than S^1) that is also a Lie group all by itself. Recall that S^1 and S^3 are the only parallelizable spheres — along with S^7. Moreover, any Lie group has a manifold whose parallelization is provided by the vector fields making up the Lie algebra that generates the Lie group. Since the direct product of certain spheres becomes parallelizable, it is as if the parallelizable nature of the submanifold S^3 infects the direct product of S^3 with the other spheres with its own parallelizability.

For future reference, we note: in the case of ADE type Coxeter graphs, the fundamental invariant polynomials of the Coxeter groups, whose degrees are the fundamental degrees, will play a very important role in the ADE classification of catastrophes (also called singularities of differentiable maps) [Arnold, 1981; 1986]. The ADE classification of simple catastrophes will be described in Chap. 7.

Appendix A: Complex Reflection Groups

In his preface to *Regular Complex Polytopes,* Coxeter wrote: "Its relationship to my earlier *Regular Polytopes* resembles that of *Through the Looking Glass* to *Alice's Adventures in Wonderland.* This book [Coxeter, 1991], which Coxeter says he "constructed ... like a Bruckner symphony," is the most beautiful mathematics book which I possess (among several hundred). Here I will be able to present only a tiny (but relevant) fraction of Coxeter's expedition through the Looking Glass.

Complex reflections are defined in complex vector spaces, $\mathbb{C}^n$, in which they must fix a complex hyperplane of dimension $n-1$. These are sometimes called pseudo reflections because, unlike reflections in real vector spaces, $\Re^n$, they are not necessarily of order 2, in which a reflection imposed twice is equivalent to the identity element. In $\mathbb{C}^n$ reflections might be of any finite order. For example a reflection of order 3 must be imposed three times to be equivalent to the identity element.

Complex reflection groups are finite groups which are generated by complex reflections. It is significant that all the Coxeter groups can be made into complex reflection groups by changing the real vector space on which they act into a complex vector space.

The complex reflection groups were completely classified by [Shephard and Todd, 1954]. They summarized the properties of these groups in a table with 37 lines. The first two lines described complex version of the Coxeter groups A_n, B_n and D_n. The third line described the cyclic groups Z_n, which we can consider as A_n-type McKay groups.

Since in this book we are especially partial to McKay groups, it is significant that the Shephard–Todd lines 4 through 22 describe complex group generalizations of the three McKay groups of E-type. The complexification of the three E-type Coxeter (Weyl) groups are described in the last three lines 35 through 37.

The Shephard–Todd groups 4 through 22 (which entail generalizations of $\mathcal{TD}, \mathcal{OD}$ and $\mathcal{ID}$) are presented in the table below, where we use Coxeter's labels for these "binary polyhedral" groups: $\mathcal{TD} = \langle 3,3,2\rangle; \mathcal{OD} = \langle 4,3,2\rangle; \mathcal{ID} = \langle 5,3,2\rangle$. The Coxeter symbol $\langle p,q,r\rangle$ refers to the exponents of the generators of these finite McKay groups.

Although these McKay groups are not themselves complex reflection groups, yet their generalizations may be such groups. In the notation used here the symbol $\langle p,q,r\rangle_s$ means that

$$\langle p,q,r\rangle = \langle p,q,r\rangle_s / Z_s,$$

where Z_s is the center. The symbol, $p[q]r \subset \langle p,q,r\rangle_s$ indicates a subgroup relationship [Coxeter, 1991].

Complex reflection groups on $\mathbb{C}^2$ of $\mathcal{TD}$ type.

S-T#: Group	**Order**	**Reflections**	**Invariants**	**Degrees**
4 3[3]3 $\subset \langle 3,3,2\rangle_6$	24	3^8	f, t	4, 6
5 3[4]3 $\subset \langle 3,3,2\rangle_6$	72	3^{16}	f^3, t	12, 6
6 3[6]2 $\subset \langle 3,3,2\rangle_6$	48	$2^6 3^8$	f, t^2	4, 12
7 $\langle 3,3,2\rangle_6$	144	$2^6 3^{16}$	f^3, t^2	12, 12

The above relative tetrahedral invariants are:

$$\mathbf{f} = x^4 + (2\sqrt{-3})x^2y^2 + y^4; \quad \mathbf{t} = xy^5 - x^5y.$$

Complex reflection groups on $\mathbb{C}^2$ of $\mathcal{OD}$ type.

8	$\mathbf{4[3]4} \subset \langle 4,3,2\rangle_{12}$	96	$2^6 4^{12}$	h, t	8, 12
9	$\mathbf{4[6]2} \subset \langle 4,3,2\rangle_{12}$	192	$2^{18} 4^{12}$	h, t^2	8, 24
10	$\mathbf{4[4]3} \subset \langle 4,3,2\rangle_{12}$	288	$2^6 3^{16} 4^{12}$	h^3, t	24, 12
11	$\boldsymbol{\langle 4,3,2\rangle_{12}}$	576	$2^{18} 3^{16} 4^{12}$	h^3, t^2	24, 24
12	**GL(2,3)** $\subset \langle 4,3,2\rangle_2$	48	2^{12}	f, h	6, 8
13	$\boldsymbol{\langle 4,3,2\rangle_2}$	96	2^{18}	f^2, h	8, 12
14	$\mathbf{3[8]2} \subset \langle 4,3,2\rangle_{12}$	144	$2^{12} 3^{16}$	f, t^2	6, 24
15	$\boldsymbol{\langle 4,3,2\rangle_6}$	288	$2^{18} 3^{16}$	f^2, t^2	12, 24

The above relative octahedral invariants are:

$$\mathbf{f} = xy^5 - x^5y; \quad \mathbf{h} = x^8 + 14x^4y^4 + y^8; \quad \mathbf{t} = x^{12} - 33x^8y^4 - 33x^4y^8 + y^{12}.$$

Complex reflection groups on $\mathbb{C}^2$ ($\mathcal{ID}$ type).

S-T#: Group		**Order**	**Reflections**	**Invariants**	**Degrees**
16	$\mathbf{5\{3\}5} \subset \langle 5,3,2\rangle_{30}$	600	5^{48}	h, t	20, 30
17	$\mathbf{5[6]2} \subset \langle 5,3,2\rangle_{\mathbf{30}}$	1200	$2^{30}5^{48}$	h, t^2	20, 60
18	$\mathbf{5[4]3} \subset \langle 5,3,2\rangle_{30}$	1800	$3^{40}5^{48}$	h^3, t	60, 30
19	$\boldsymbol{\langle 5,3,2\rangle}_{30}$	3600	$2^{30}3^{40}5^{48}$	h^3, t^2	60, 60
20	$\mathbf{3[5]3} \subset \langle 5,3,2\rangle_{30}$	360	3^{40}	f, t	12, 30
21	$\mathbf{3[10]2} \subset \langle 5,3,2\rangle_{30}$	720	$2^{30}3^{40}$	f, t^2	12, 60
22	$\boldsymbol{\langle 5,3,2\rangle_2}$	240	2^{30}	f, h	12, 20

The above absolute icosahedral invariants are:

$$\begin{aligned}
\mathbf{f} &= xy(x^{10} - y^{10} + 11x^5y^5),\\
\mathbf{h} &= 228x^5y^5(x^{10} - y^{10}) - (x^{10} + y^{10}) - 494x^{10}y^{10},\\
\mathbf{t} &= x^{30} + y^{30} + 522x^5y^5(x^{20} - y^{20}) - 10005x^{10}y^{10}(x^{10} - y^{10}).
\end{aligned}$$

Note: In general, the product of the degrees is the order of the group. Also note that **h** is the Hessian of **f**; and **t** is the Jacobian of **h** and **f**. Moreover, the following syzygies hold for the absolute invariants:

$\mathcal{TD}$: $t^2 = h^3 - 108f^4$ (where t, h and f are also the relative octahedral invariants.

$\mathcal{OD}$: $Z^2 = Y(X^3 - 108Y^2)$, where $X = h, Y = f^2$, and $Z = ft$.

$\mathcal{ID}$: $t^2 = -h^3 + 1728f^5$.

For $\mathcal{TD}$: $f = 0 \rightarrow$ six vertices of the octahedron.

$h = 0 \rightarrow$ eight vertices of the cube.

$t = 0 \rightarrow 12$ vertices of the cuboctahedron.

For $\mathcal{ID}$: $f = 0 \rightarrow 12$ vertices of the icosahedron.

$h = 0 \rightarrow 20$ vertices of the dodecahedron.

$t = 0 \rightarrow 30$ midpoints of the edge lines of the dodecahedron.

It must be understood that these polyhedral designations are based on viewing these formulas as complex functions on the Argand plane for $\mathbb{C}^1$. The polyhedral vertices as points on S^2 are conformally mapped to $\mathbb{C}^1$.

For example: The $\mathcal{TD}$ absolute invariant, $f = xy^5 - x^5y = 0$, has the six-point solution set $\{0, \infty, \pm 1, \pm i\}$. These are complex numbers mapped to S^2, with 0 at the origin of $\mathbb{C}^1$ (which is the "south pole" of S^2); ∞ at the "north pole;" while ± 1 are $\pm$i are mapped to the four "cardinal" points of the "equator" of S^2. Clearly these six points are the vertices of an octahedron embedded in S^2 [Klein, 1956, p. 54].

See also [Shephard and Todd, 1954; Duval, 1964; Coxeter and Moser, 1965; Coxeter, 1991].

Appendix B: Folding ADE Graphs

$^{\infty}\mathrm{D}_{n+2}$ o–o–o–•••–o–o–o (with a node o hanging from the second and from the second-to-last node) $\Rightarrow$ $^{\infty}\mathrm{B}_n$ o–4–o–o–•••–o–o–4–o

(fold the end nodes together)

$^{\infty}\mathrm{D}_{2n}$ o–o–o–•••–o–o–o (with a node o hanging from the second and from the second-to-last node) $\Rightarrow$ $^{\infty}\mathrm{C}_n$ o–4–o–o–•••–o–o–o (with a node o hanging from the second-to-last node)

(fold the left nodes over the right nodes)

$^{\infty}E_6$ o–o–o–o–o (with two nodes hanging below the middle node: o, o) $\Rightarrow$ $^{\infty}G_2$ o–6–o–o

(fold all three arms together)

$^{\infty}E_7$ o–o–o–o–o–o–o (with one node hanging below the middle node: o) $\Rightarrow$ $^{\infty}F_4$ o–o–4–o–o–o

(fold the left arm over the right arm and swing the dangling arm to the horizontal position) see [Hiller, 1982, p. 35].

Chapter 7

Thom–Arnold Catastrophe Structures

Catastrophe theory was initially developed by Rene Thom in the '60s in order to provide a mathematical basis for dynamical systems which can undergo extremely rapid change while the parameters controlling the system change only slightly. This theory is closely related to bifurcation theory and singularity theory, and Thom used and further developed techniques from these theories [Thom, 1975].

The dynamical systems that inspired Thom's early work were from embryology and he hoped to provide a mathematical framework for biological development. He was especially influenced by Thompson's highly illustrated book *On Growth and Form* [Thompson, 1942], and Waddington's *The Strategy of the Genes: A Discussion of Some Aspects of Theoretical Biology* [Waddington, 1957].

In particular, Thom drew mathematical ideas from Waddington's concept of an "epigenetic landscape." This was a geometric model for the transition from genotype to phenotype (i.e. from the genetic information to the biological form). Waddington's epigenetic landscape was a 2D surface like that of a valley with many ramifying branches, which he called *chreods*. A ball rolling downhill through this valley under the influence of the gravitational potential has many different possible paths to the lower levels.

In Thom's geometrization of these ideas, the gravitational potential is replaced by the potential functions of his catastrophe structures. The chreods become aspects of the stability required by these structures. Thom's structural stability means that a small perturbation of the structure leaves singularities unchanged. These singularities correspond to the minima of the potential function.

An especially fruitful Thomian idea was the attractor and basins of attractors, which had been inspired by Waddington's chreods.

Each singularity would be associated with an attractor. Moreover, an attractor basin contained the points of the state space that would inevitably migrate into the attractor. The various attractor basins would be separated from each other by a well-defined structure that Thom called the discriminant, and is also called the separatrix [Gilmore, 1981].

Thom's attractor idea was later developed, by other mathematicians, into the strange attractors of chaos theory. These strange attractors are characterized by necessarily being fractal, as in *The Fractal Geometry of Nature* [Mandelbrot, 1983]. Chaos theory can be considered as a specialized part of catastrophe theory [Arnold, 1986; Gilmore, 1981]. Here, however we will limit our focus on the simple catastrophes, since they were ADE classified by Arnold.

Thom limited the control space of his catastrophe systems to be 4D, because he wanted to apply his theory to the biological world in spacetime. He conjectured that with control spaces of dimension less than or equal to four, there are only seven stable catastrophe structures. He listed them as follows [Thom, 1975]:

Fold: $V = x^3 + ax$

Cusp: $V = x^4 + ax + bx^2$

Swallowtail: $V = x^5 + ax^3 + bx^2 + cx$

Butterfly: $V = x^6 + ax + bx^2 + cx^3 + dx^4$

Hyperbolic umbilic: $V = x^3 + y^3 + axy - bx - cy$

Elliptic umbilic: $V = x^3 - 3xy^2 + a(x^2 + y^2) - bx - cy$

Parabolic umbilic: $V = x^2y + y^4 + ax^2 + by^2 - cx - dy$

In these formulas, V is a potential function, made up of the germ followed by the control parameter terms (for a versal unfolding) with coordinates $\{a, b, c, d\}$. Thom's various names for these seven catastrophes suggest the form of the bifurcation in the control space. In this sense, catastrophe theory can be considered as a branch of the very active work in bifurcation theory [Golubitsky and Schaeffer, 1985].

Similarly, catastrophe theory is a specialization of the very broad field of singularity theory, which Thom drew upon in order to propose his conjecture about the stability of mappings between manifolds of dimension n to dimension $n+p$, where $n \leq 2$, and $p \leq 4$.

This Thom conjecture was turned into a theorem by Mather, who proved the stability of these smooth mappings [Golubitsky and Guillemin, 1973].

Moreover, Arnold, who had previously done much pioneering work in singularity theory, was able to embed Thom's seven elementary catastrophes in an infinite set of simple catastrophes. These simple catastrophes were classified by the ADE Lie algebra labels, in which the Lie algebra rank is $r = p+1$, where p is the dimension of the control parameter space of the simple catastrophe. We can list Arnold's classification of simple catastrophes as follows:

A_2: **Fold:** $V = x^3 + ax$
A_3: **Cusp:** $V = x^4 + ax + bx^2$
A_4: **Swallowtail:** $V = x^5 + ax + bx^2 + cx^3$
A_5: **Butterfly:** $V = x^6 + ax + bx^2 + cx^3 + dx^4$
A_r: $V = x^{r+1} + y^2 + a_1x + \cdots + a_nx^{r-1}$
D_4^+: **Hyperbolic umbilic:** $V = x^2y + y^3 + x + by + cy^2$
D_4^-: **Elliptic umbilic:** $V = x^2y - y^3 + ax + by + cy^2$
D_5: **Parabolic umbilic:** $V = x^2y + y^4 + ay + by^2 + cxy + dxy^2$
D_r: $V = x^2y + y^{r-1} + a_1y + \cdots + a_ny^{r-2} + bx$
E_6: $V = x^3 + y^4 + ay + by^2 + cx + dxy + exy^2$
E_7: $V = x^3 + xy^3 + ay + by^2 + cy^3 + dy^4 + exy + fx$
E_8: $V = x^3 + y^5 + ax + by + cxy + dy^2 + ey^3 + fxy^2 + gxy^3$

[Thom, 1975; Arnold, 1981; 1986; Gilmore, 1981].

Note that the general A_r germ has an extra y^2 term. This accords with the fact that, for any germ, we can add any number of quadratic terms to the germ without altering the bifurcation structure [Bruce and Giblin, 1984, p. 201].

In order to clarify catastrophe structures, it is useful to view a diagram of the A_3 catastrophe manifold in $\mathfrak{R}^3$. Here we picture the real catastrophe manifold which is the set of critical points of the A_3 (cusp type) catastrophe polynomial:

$$V = x^4 + ax + bx^2 \text{ (cf. p. 66)}$$

rewritten as:

$$V = A^4 + t_1 A^2 + t_2 A.$$

The catastrophe manifold is projected down to the control space (here 2D) and the folds of the manifold become cusp lines in the control space. Such a control space diagram is called a bifurcation. The same bifurcation occurs if we add any number of quadratic terms to V.

The A_3 catastrophe manifold in $\mathfrak{R}^3$ is the zero-set of

$$dV/dA = 4A^3 + 2t_1 A + t_2;$$

and the bifurcation diagram in $\mathfrak{R}^2$ is the zero-set of

$$d^2V/dA = 12A^2 + 2t_1.$$

Note that the bifurcation diagram has a cusp singularity.

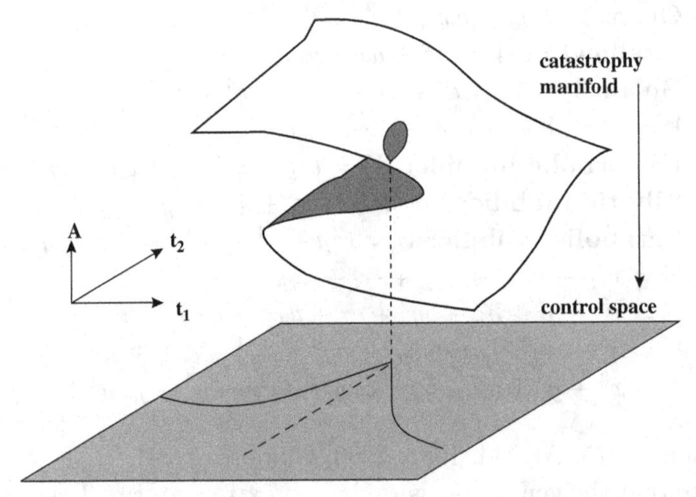

A_3 catastrophe manifold and bifurcation diagram.

The A_3 catastrophe structure is also pictured via the real critical value surface embedded in the 3D base space $\mathfrak{R}^3$ (with parameters t_1, t_2, t_3) of the A_3 catastrophe bundle B^5, which is the zero-set of the A_3 polynomial

$$V = A^4 + B^2 + C^2 + t_1 A^2 + t_2 A + t_3.$$

The A_3 critical value surface is the value of V on the catastrophe manifold of critical points (as in the figure above).

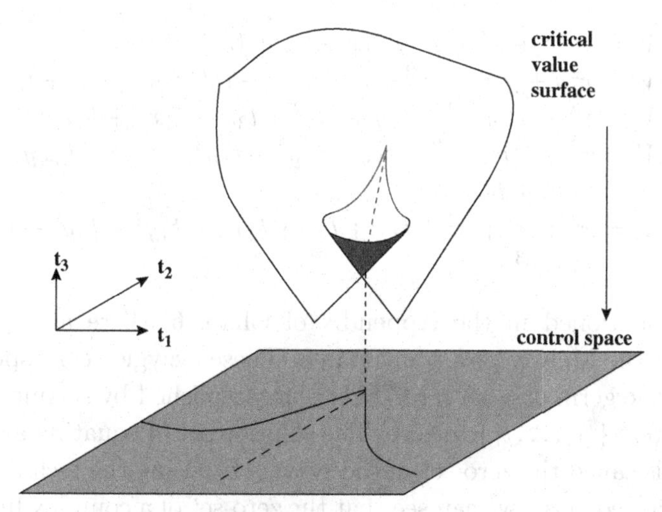

A_3 critical value surface in $\Re^3$ (with parameters t_1, t_2, t_3).

In order to clarify the relationships between these catastrophe mappings and the ADE Lie algebra labels, it is necessary to consider them as complex mappings.

As is well known, the complex version of a theory usually simplifies the theory. Indeed, in this case, for example, The D_4^+ and D_4^- mappings become real parts of the complex D_4 mapping.

Also in Arnold's classification scheme, the only catastrophes which lack differing real versions are: A_r (where r is even), E_7 and E_8 [Gilmore, 1981, p. 138].

This is reminiscent of the fact that the classification of Lie algebras is primarily the classification of complex Lie algebras, with various real forms (such as the compact real form) corresponding to each complex version [Gilmore, 1974].

Moreover, we make use of an added quadratic term in each germ to accommodate the absolute invariants of the McKay group $\mathcal{m}$ associated with the ADE Lie algebras. Also note that we have written the parameters as t_n (where n runs from 1 through the rank, r). These parameters are the invariants of the Coxeter–Weyl groups associated with the ADE Lie algebras. So these catastrophe mappings from $\mathbb{C}^{r+3}$ into $\mathbb{C}^3$ are written entirely as mappings of invariant polynomials $\{x, y, z, t_1, \ldots t_r\}$ [Gilmore, 1981, p. 49].

$$
\begin{aligned}
\mathrm{A}_r: \quad & V = x^{r+1} + y^2 + z^2 + t_1 x + \cdots + t_r x^{r-1} \\
\mathrm{D}_r: \quad & V = x^2 y + y^{r-1} + z^2 + t_1 y + \cdots + t y^{r-2} + t_{r-1} x + t_r \\
\mathrm{E}_6: \quad & V = x^3 + y^4 + z^2 + t_1 y + t_2 y^2 + t_3 x + t_4 xy + t_5 xy^2 + t_6 \\
\mathrm{E}_7: \quad & V = x^3 + xy^3 + z^2 + t_1 y + t_2 y^2 + t_3 y^3 + t_4 y^4 + t_5 xy \\
& \quad + t_6 x + t_7 \\
\mathrm{E}_8: \quad & V = x^3 + y^5 + z^2 + t_1 x + t_2 y + t_3 xy + t_4 y^2 + t_5 y^3 + t_6 xy^2 \\
& \quad + t_7 xy^3 + t_8.
\end{aligned}
$$

As mentioned in the Appendix of Chap. 6, there are syzygies between the McKay group invariants. These syzygies correspond to setting the germ of V to zero. This is accomplished by setting all the parameters $\{t_1, \ldots, t_r\}$ to zero. The solution set of equating a syzygy to zero is called the zero-set of the syzygy. By using the techniques of algebraic geometry we can see that the zero-set of a complex function of n variables will be an $(n-1)$-dimensional complex subspace in an n-dimensional complex space [Kendig, 1977].

Thus in the case of these ADE syzygies, the zero-set of a germ is a complex 2D subspace in complex 3D space with coordinates $\{x, y, z\}$. A useful analogy is the real-space formula for a parabola, ordinarily written as $y = x^2$, but in algebraic-geometry format as:

$$X^2 - Y = 0.$$

Therefore the parabola, as a zero-set, is seen as a 1D subspace in the 2D (X, Y)-vector space. The parabola is a very simple example of a singularity space, since the derivative dy/dx of the parabola at the origin of the (X, Y)-axis is 0 and is thus a critical point. Note also that the parabola itself is a Whitney fold and is a local potential in the unfoldings of several Thom catastrophes, most vividly in the Cusp and Swallowtail [Thom, 1975; Arnold, 1981; Arnold *et al.*, 1985].

Moreover, because the ADE syzygies are polynomials of McKay group invariants, it can be shown that these zero-sets are hypersurface singularity structures, which in each case is the quotient space:

$$\mathbb{C}^2/\mathcal{m} \text{ (where } \mathcal{m} \text{ is a McKay group).}$$

This corresponds to the fact that the McKay groups are finite subgroups of $\mathcal{SU}(2)$, while this Lie group acts (by definition) as a unitary transformation group on $\mathbb{C}^2$.

Incidentally, because these syzygies were first described by Felix Klein in his book *Lectures on the Icosahedron* [Klein, 1884; 1913], these singularities are called Kleinian singularities [Looijenga, 1984]. The Kleinian singularities are simple in the sense that they exist as a single cusp-like point at the origin of $\mathbb{C}^3$.

In the terminology of singularity theory and catastrophe theory, a singularity is an equivalence class of a germ at a critical point (i.e. a point where all the partial derivatives vanish). A space with such a singularity is not everywhere differentiable, and is called by the more general term variety. Especially important are complex varieties [Kendig, 1977; Gilmore, 1981; Arnold *et al.*, 1985].

Given a Kleinian singularity space $\mathbb{C}^2/m$, there are three ways to transform it into smoother spaces, and ultimately into a smooth manifold.

(1) **Resolution**: lift $\mathbb{C}^2/m$ into a higher dimensional space so that the singularity of the lower dimensional space can be viewed as a projection from a smooth manifold. A simple real-space illustration of this would be the shadow in 2D space of a smoothly curved wire in 3D space. The shadow seen from a certain angle could easily have a kink — a "cusp" singularity. This lifting of the singularity structure is called a resolution or a desingularization, and will be described more fully in Chap. 8 on the ADE classification of ALE spaces.
(2) **Topologize:** intersect $\mathbb{C}^2/m$ in the neighborhood of the singularity by surrounding it with a small sphere S^5 in $\mathfrak{R}^6(=\mathbb{C}^3)$. This leads to knot theory to be described in Chap. 9.
(3) **Unfolding:** deform $\mathbb{C}^2/m$ in the fiberspace, whose base space is the corresponding catastrophe parameter space. In this case $\mathbb{C}^2/m$ is the identity fiber at the origin of the base space. This is the desingularization procedure to be described here.

The McKay correspondence is a deep relationship between the finite subgroups of $\mathcal{SU}(2)$ and the affine Lie algebras (cf. Chap. 4). Here, however, we will be describing a relationship between the subgroups of $\mathcal{SU}(2)$ and the finite Coxter groups classified by the ADE simple Lie algebras. This is because the parameter space for the unfolding of the catastrophe germ corresponds exactly to the ADE Coxeter graphs. This implies that the basis in the unfolding space consists of the polynomial invariants of the Coxeter group (as mentioned above). So we can consider the Coxeter graphs as unfolding diagrams, which are also called intersection diagrams [Looijenga, 1984]. The germs are listed here, while the germs with unfolding formulas were listed previously in this chapter.

Unfolding graph		McKay group	Germ of ADE catastrophe	Degrees of invariants x, y, z
A_r	o–o–. . . –o	$\mathcal{Z}_{r+1}$	$x^{r+1} + y^2 + z^2$	$2, r+1, r+1$
D_r	o–o–. . . –o (branch o on 2nd node)	$\mathcal{Q}_{r-2}$	$x^2 y + y^{r-1} + z^2$	$4, 2(r-2), 2(r-1)$
E_6	o–o–o–o–o (branch o on 3rd node)	$\mathcal{TD}$	$x^3 + y^4 + z^2$	6, 8, 12
E_7	o–o–o–o–o–o (branch o on 3rd node)	$\mathcal{OD}$	$x^3 + xy^3 + z^2$	8, 12, 18
E_8	o–o–o–o–o–o–o (branch o on 3rd node)	$\mathcal{ID}$	$x^3 + y^5 + z^2$	12, 20, 30

Note: The degrees of the invariant polynomials $\{x, y, z\}$, are considered as "weights" of the catastrophe germ, viewed as a weighted homogeneous polynomial. The overall degree of this polynomial is, in each case, two times the highest invariant degree.

For example in the case of E_7, we have for the $\mathcal{OD}$ germ:

$$X^3 + XY^3 + Z^2$$

where X, Y and Z are $\mathcal{OD}$ invariants of degrees 12, 8 and 18, so that:

$$3(12) = 12 + 3(8) = 2(18) = 36.$$

Cf. p. 63, where we have written the $\mathcal{OD}$ syzygy as:

$$Z^2 = X^3 Y - 108Y^3.$$

The absolute octahedral invariants, X, Y and Z, can be unpacked via the relative octahedral invariants, f, h and t:

$$X = h; \quad Y = f^2; \quad Z = ft,$$

where $f = xy^5 - x^5y$;

$$h = x^8 + 14x^4y^4 + y^8;$$
$$t = x^{12} - 33x^8y^4 - 33x^4y^8 + y^{12}$$

[Klein, 1956; Duval, 1964; Steinberg, 1959].

Moreover, since the catastrophe mappings (p. 67) entail both McKay group and Coxeter group invariants, it is not too surprising to find a relationship between the degrees of the McKay group invariants and the corresponding Coxeter reflection group. This relationship identifies the Coxeter number with the highest degree of the corresponding McKay group invariant. For example:

12 = E_6 Coxeter number = degree of the highest $\mathcal{TD}$ invariant.

18 = E_7 Coxeter number = degree of the highest $\mathcal{OD}$ invariant.

30 = E_8 Coxeter number = degree of the highest $\mathcal{ID}$ invariant.

The use of Coxeter graphs as unfolding diagrams implies that the number of parameters for the unfolding is the rank of the graph. But much more is implied. The Coxeter graphs describe the reflection hyperplanes — one basic mirror to each node on the graph. The reflections in these basic mirrors generate the full set of mirrors, which form a Coxeter arrangement of hyperplanes embedded in the complex vector space $\mathbb{C}^r$, where r is the rank of the Coxeter graph. This arrangement consists of a set of n chambers, where n is the order of the Coxeter group.

Analogous to the action of the McKay group $\boldsymbol{m}$ on $\mathbb{C}^2$, which yields the quotient structure $\mathbb{C}^2/\boldsymbol{m} \subset \mathbb{C}^3$, the Coxeter group W acts on $\mathbb{C}^r$ to create the quotient space $\mathbb{C}^r/W$.

The action of the Coxeter reflection group W on $\mathbb{C}^r$ is very intricate. By various combinations of reflections W creates reflection orbits in $\mathbb{C}^r$. If the number of points in an orbit is exactly the order W, we call the orbit a regular orbit. Such orbits move points from chamber to chamber, between the mirror hyperplanes. Since the number of chambers is the order of W, the cardinality of such a regular orbit must be the same as the order of W. If, however, an orbit of W consists of points within the mirrors themselves, the orbit would be limited to the number of mirrors, which is less than the number of chambers. Such orbits (jumping from mirror to mirror) are called special orbits.

We note that each point of $\mathbb{C}^r/W$ is an orbit in $\mathbb{C}^r$ so that $\mathbb{C}^r/W$ is an orbit space and is called an orbifold. Thus the points in $\mathbb{C}^r/W$ are of two types: regular and special, corresponding to whether they are regular or special orbits in $\mathbb{C}^r$. Since hyperplanes are $(n-1)$-dimensional planes in an n-dimensional space, the wonderful consequence is that the Coxeter arrangement of hyperplane mirrors in $\mathbb{C}^r$ is transformed into a hyperspace of dimension $r-1$ in $\mathbb{C}^r/W$. This hyperspace, made up of special orbits in $\mathbb{C}^r$, is indeed very special. This hyperspace is the separatrix Σ for the unfolding structure of the catastrophe. This is a very complicated structure with hyperspace walls intersecting each other and creating chambers in $\mathbb{C}^r/W$. The separatrix is also called the critical value surface, such as in the A_3 (real subspace) diagram above.

Since $\mathbb{C}^r/W$ is the base space of the catastrophe fiber space, the identity fiber at the origin of $\mathbb{C}^r/W$ must be the singularity variety $\mathbb{C}^2/\mathfrak{m}$. This is because $\mathbb{C}^2/\mathfrak{m}$ is the zero set of the catastrophe germ, and it is specified by setting all the unfolding parameters $\{t_1, \ldots, t_r\}$ to zero. As one moves away from the origin of $\mathbb{C}^r/W$ by changing the values of the parameters, different fiber versions of the identity are selected. In general the farther the fiber is from the origin, the milder is the singularity structure of the fiber. This is the sense in which $\mathbb{C}^2/\mathfrak{m}$ becomes unfolded.

Moreover, moving the fiber point away from the origin along any path in $\mathbb{C}^r/W$ means that various hypersurfaces of Σ must be crossed. Each chamber of Σ bounds an open set of regular points attached to versions of the fiber $\mathbb{C}^2/\mathfrak{m}$ which are topologically equivalent. Thus within any chamber of Σ the changes in the fiber are not abrupt, but gradual. However, when a fiber path crosses Σ, the fiber changes abruptly. Since Σ is a hypersurface (of dimension $r-1$) it is thin within the ambient space $\mathbb{C}^r/W$. This is why the fiber, while crossing the separatrix Σ, can undergo a drastic transformation dependent only on a tiny parameter change. Thus the term catastrophe seems appropriate for these ADE structures.

The catastrophe fiber bundle for any ADE-type catastrophe is simply the zero set of the full deformation mapping form. For example the A_3 mapping form is:

$$V = x^4 + y^2 + z^2 + t_1 x^2 + t_2 x + t_3.$$

These are complex variables. However, in a real subspace $\mathfrak{R}^3$, the most relevant catastrophe structure can be pictured with a diagram, in which t_1 and t_2 are real horizontal parameters, while t_3 is a real vertical parameter of $\mathfrak{R}^3$. Since the three parameters $\{t_1, t_2, t_3\}$ are invariants of W, we find embedded in $\mathfrak{R}^3$ the separatrix Σ, which is also equivalent to the critical value surface of the A_3 catastrophe (cf. the diagram on p. 69).

A more complicated example is the E_7 deformation mapping form:

$$V = x^3 + xy^3 + z^2 + t_1 y + t_2 y^2 + t_3 y^3 + t_4 y^4 + t_5 xy + t_6 x + t_7.$$

The 6D separatrix Σ is in $\mathbb{C}^7/W(E_7)$, since $\{t_1, \ldots, t_7\}$ are invariants of the E_7 Coxeter group (also called the Weyl group). Since there are 63 mirror hyperplanes in $\mathbb{C}^7$, 63 is the maximum number of points in the special orbits making up Σ in $\mathbb{C}^7/W(E_7)$. By contrast, the regular orbits have $288 \times 7! \times 2 = 2{,}903{,}040$ elements, which is the order of $W(E_7)$. These regular orbits are the points inside the chambers between the 6D walls of Σ.

Note that 288 is the product of the E_7 balance numbers (1, 2, 3, 4, 3, 2, 1, 2), while 7! is 5040, the order of the Symmetric-7 group which permutes the seven basic mirrors of E_7; and 2 is the order of $\mathcal{Z}_2$ which acts on the bilateral symmetry of the extended E_7 Coxeter graph. Since the E_7 balance numbers are also (via the McKay correspondence) the iirep dimensions of $\mathcal{OD}$, this is yet another way in which the structure of the McKay group and the Coxeter group intertwine.

In general, for any ADE Lie algebra of rank r:

(1) The control parameters of the catastrophe bundle are $\{t_1, \ldots, t_{r-1}\}$.

(2) The t_r parameter (always with 1 as coefficient) plays the role of time along the many paths ramifying out from the origin of $\mathbb{C}^r/W$, where there is attached the identity fiber $\mathbb{C}^2/m$ (where m is a McKay group).

(3) Movement along any of these paths corresponds to the selection of different values of the control parameters, and thus different fibers which entail an unfolding of the singularity structure $\mathbb{C}^2/m$.

(4) The changes in the fiber attached to a path are mild if the movement along the path (while picking out different fibers) remains within a chamber of the separatrix. However, if movement along the path crosses the separatrix, the change in the fiber will be drastic.

(5) As fibers farther and farther from the origin in $\mathbb{C}^7/W$ are encountered and more and more separatrix walls are crossed, the fibers become more and more unfolded [Arnold, 1981, 1986; Arnold *et al.*, 1985].

Beyond the unfoldings of the fibers as described above, the ADE Lie algebras themselves provide the structures for the resolution of the deformed (unfolded) fibers, including most importantly the identity fiber $\mathbb{C}^2/m$ [Brieskorn, 1970; Tyurina, 1970; Slodowy, 1983].

This unfolding can be clarified by the projection diagram:

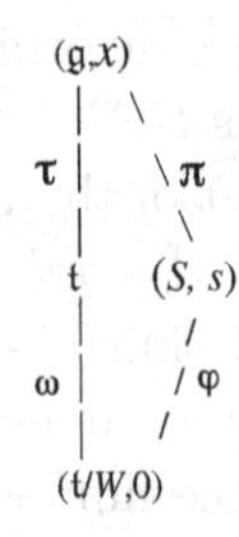

where $\mathfrak{g}$ is an ADE-type Lie algebra, and x is a subregular nilpotent element within the nilpotent variety in $\mathfrak{g}$. The Lie group version of this projection diagram takes G to its maximal torus T and then to T/W.

The projection τ is from $\mathfrak{g}$ onto its Cartan subalgebra:

$$\mathfrak{t}(= \mathbb{C}^r, \text{ where } r \text{ is the rank of } \mathfrak{g}).$$

The projection ω is from $\mathfrak{t}$ to the orbit space $\mathfrak{t}/W$, where 0 is the origin point in $\mathfrak{t}/W$.

The projection π is from $\mathfrak{g}$ to S, which is the $(r+2)$-dimensional slice, which is transverse to the nilpotent variety in $\mathfrak{g}$; and s is a subregular (i.e. singular) element in this variety.

The nilpotent variety $\mathfrak{n}$ in $\mathfrak{g}$ is the identity fiber in the fiber bundle with projection:

$$\chi\colon (\mathfrak{g}, x) \longrightarrow (\mathfrak{t}/W, 0)$$

so that the dimensionality of $\mathfrak{n}$ is:

$$\text{dim.}(\mathfrak{g}) - \text{dim.}(\mathfrak{t}/W) = (\text{Cox\#})(\text{Rank}).$$

Note: The nilpotent variety $\mathfrak{n}$ in $\mathfrak{g}$ is the set of elements g of $\mathfrak{g}$ which via a series of adjoint multiplications $[X, Y]$ in $\mathfrak{g}$ eventually arrives at 0. That is $(\text{ad}X)^n = 0$ for some sufficiently large n.

The projection φ maps the ADE (Kleinian) singularity $\mathbb{C}^2/\mathfrak{m}$ onto the origin point 0 of $\mathfrak{t}/W$, and as a universal deformation maps unfolded versions of $\mathbb{C}^2/\mathfrak{m}$ onto the parameters $\{t_1, \ldots, t_r\}$ which are homogeneous polynomial invariants of the corresponding ADE Coxeter–Weyl group.

While φ provides for the deformation (or unfolding) of the Kleinian singularities, it must be the lifting of the slice S into the nilpotent variety $\mathfrak{n}$ that provides for the simultaneous resolution (or desingularization) of all the fibers in S.

In particular, the most singular fiber is the identity fiber in S, which is the Kleinian singularity structure $\mathbb{C}^2/\mathfrak{m}$. In the process of desingularization, the singular point evolves into a series of exceptional curves, which are 1D complex projective lines $\mathfrak{P}^1$, which geometrically are a "bouquet" of 2D spheres. Amazingly, this bouquet takes the form of a dual structure to the appropriate ADE Coxeter graph.

For example: in the E_7 case, the Kleinian singularity $\mathbb{C}^2/\mathcal{OD}$, has its singular point resolved into a bouquet of 7 (2D)-spheres:

OO8OOO

in which the spheres touch one another in a pattern which is exactly dual to the E_7 Coxeter graph

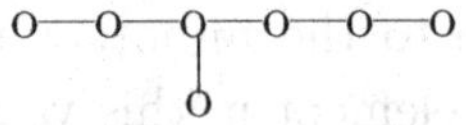

```
o—o—o—o—o—o
    |
    o
```

However, there is much more to say about the resolution of Kleinian singularities, which will be described in the following chapter.

Appendix on Catastrophe Abutments

One thing that has not been emphasized in this chapter is the fact that although the catastrophe germs are not embedded within each other, there is an embedding of the ADE parameters, and thus an embedding of the singularities and their unfoldings.

This embedding is entirely determined by the Coxeter graphs, accomplished by removing nodes from the Coxeter graphs. The rule is that for any Coxeter graphs Γ and Γ' (in which Γ' can be derived from Γ by removing a node), the unfolding of catastrophe Γ' is embedded in the unfolding of catastrophe Γ. This embedding is called an abutment.

The embeddings also correspond to the subalgebra structures of the ADE Lie algebras.

For example E_7 has E_6, D_6 and A_6 as subalgebras:

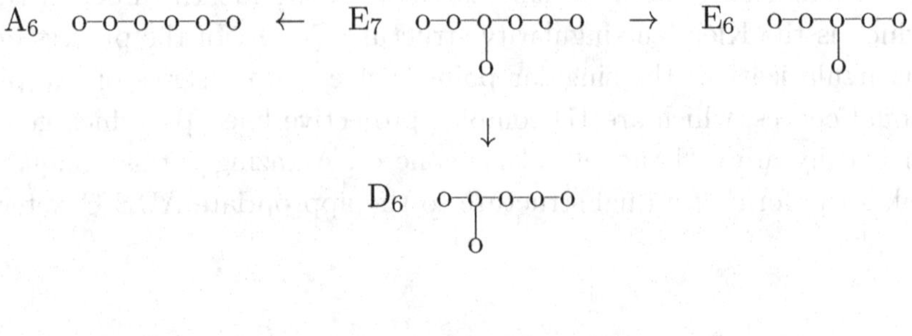

```
A6  o-o-o-o-o-o  ←  E7  o-o-o-o-o-o  →  E6  o-o-o-o-o
                            |                  |
                            o                  o

                            ↓

                    D6  o-o-o-o-o
                          |
                          o
```

Thus the set of the embeddings can be organized by the following diagram. A complete set of embeddings would include diagonal arrow-lines, such as from E_7 to A_6. Also these abutments are for complex catastrophes. The abutment diagram for the real catastrophes (contained in the complex structures) is more complicated [Gilmore, 1981, pp. 138–139].

Catastrophe abutment and Lie subalgebra diagram.

$$\begin{array}{ccccccccccccccccc}
A_1 & \leftarrow & A_2 & \leftarrow & A_3 & \leftarrow & A_4 & \leftarrow & A_5 & \leftarrow & A_6 & \leftarrow & A_7 & \leftarrow & A_8 & \leftarrow \\
 & & & & \uparrow & & \uparrow & & \uparrow & & \uparrow & & \uparrow & & \uparrow & \\
 & & & & D_4 & \leftarrow & D_5 & \leftarrow & D_6 & \leftarrow & D_7 & \leftarrow & D_8 & \leftarrow & D_9 & \leftarrow \\
 & & & & & & & & \uparrow & & \uparrow & & \uparrow & & & \\
 & & & & & & & & E_6 & \leftarrow & E_7 & \leftarrow & E_8 & & & \\
 & & & & & & & & \uparrow & & \uparrow & & \uparrow & & & \\
 & & & & & & & & T_{333} & & T_{244} & & T_{236} & & &
\end{array}$$

Note that the D series begins with D_4 because D_3 is the same as A_3, which is clear from the D_3 graph:

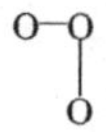

o–o which is equivalent to o–o–o.
   |
   o

Note especially that the three E structures provide gateways into the non-simple catastrophe realm, which ramifies vastly beyond the simple ADE structures [Arnold, 1981; 1986; Arnold *et al.*, 1985; Gilmore, 1981, pp. 451–457].

Chapter 8

ALE Spaces and Gravitational Instantons

The resolution of the ADE singularity structure $\mathbb{C}^2/\mathfrak{m}$, where $\mathfrak{m}$ is a finite subgroup of $\mathcal{SU}(2)$, is accomplished by the lifting of $\mathbb{C}^2/\mathfrak{m}$ into a higher dimensional space $\mathbb{C}^n$. This lifting has already been described in Chap. 7 as a key part of the universal resolution of the unfolding of $\mathbb{C}^2/\mathfrak{m}$.

However, in this chapter we will be concentrating on the desingularization of $\mathbb{C}^2/\mathfrak{m}$ itself and its description as a gravitational instanton. This entails the ADE classification of asymptotically locally Euclidean (ALE) spaces.

Moreover, there is a surprising connection between $\mathbb{C}^2/\mathfrak{m}$ and supergravity and superstring theory. In fact, superconformal quantum field theories have superpotentials, which are the germs of the ADE catastrophe structures (as described in Chap. 7). Thus $\mathbb{C}^2/\mathfrak{m}$ is the zero-set of a superpotential of a superconformal quantum field theory.

These rather striking claims require much unpacking, which will be the substance of this chapter.

A plausible starting point is the fact that string theories are described as 2D worldsheets swept out by the string and embedded in 10D spacetime. It is very significant that 2D conformal field theories (CFTs) are ADE classified [Kaku, 1999; Henkel, 1999].

Moreover, the parameters (σ, τ) of the worldsheet obey several symmetries: reparametrization invariance, 2D Lorentz invariance, supersymmetry, and conformal invariance.

Conformal invariance means that conformal mappings, changing scale but not directions, are allowed. Such changes do not affect the physical interpretation of the theory.

There are five known superstring theories; and they have all been unified by dualities and embedding in an 11D theory called M-theory, which in turn is embedded in a 12D theory called F-theory. Edward Witten, who discovered M-theory [Witten, 1995], says that M stands for "magic, mystery or membrane — according to taste." However, since Cumrun Vafa discovered F-theory [Vafa, 1996], M and F may stand for Mother and Father.

We can outline the duality relationships between the various theories by the following diagram, with dimensions (12), (11) and (10):

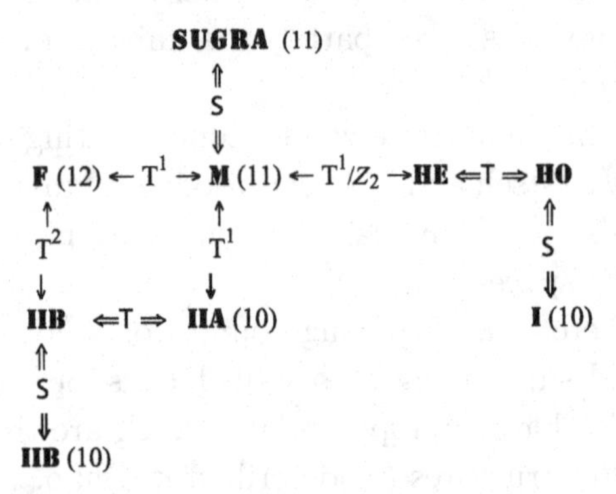

Superstring theory types, dualities and embeddings.

Note: T^1 is a 1D Torus ($= S^1$, circle); T^2 is a 2D Torus ($= S^1 \otimes S^1$); T^1/Z_2 is a line segment (Z_2 is the cyclic-2 group $\{\pm 1\}$).

$\mathcal{T}$ is T-duality. It exchanges compactifications on large and small Tori, by exchanging size for string winding number.

$\mathcal{S}$ is S-duality. It exchanges weak and strong effects of gauge coupling constants. Note that **IIB** is self-dual. Also this duality makes 11D supergravity (SUGRA) a "low energy" limit of M-theory.

Type **I** has both open and closed strings; and reduces to $N = 1$ SUGRA coupled to $\mathcal{SO}(32)$ gauge multiplets.

Type **IIA** has closed strings (right movers and left movers, which rotate in an opposite sense); and reduces to $N = 2$ (non-chiral) SUGRA.

Type **IIB** has closed strings (right movers and left movers, which rotate in an opposite sense); it reduces to $N = 2$ (chiral) SUGRA.

Type **HE** is Heterotic $E_8 \otimes E_8$, and has closed strings, which are of two types: right-movers with 10D supersymmetry and left-movers in the bosonic 26D spacetime; it reduces to $N = 1$ SUGRA coupled to $E_8 \otimes E_8$ gauge multiplets.

Type **HO** is Heterotic $\mathcal{SO}(32)$ and has closed strings of two types exactly like those of **HE**; it reduces to $N = 1$ SUGRA coupled to $\mathcal{SO}(32)$ gauge multiplets.

SUGRA (11) is 11D supergravity, which is a point-particle theory (rather than a string theory). It is the maximally extended form of supergravity with eight supersymmetry transformations. Like all versions of supergravity, it contains the graviton (a massless spin-2 particle) by necessity, and thus has been a strong candidate for a quantum gravity theory. But for many reasons it is necessary to embed this theory in the richer world of M-theory and its family of superstring theories.

M-theory (11D) is a work in progress. It is neither a point particle nor string theory. Its elementary structures are M2-branes and M5-branes.

F-theory (12D) is also work in progress. It originated as an extension of **IIB** theory with a 2D Torus parametrizing two scalar fields, the dilaton and the R–R scalar.

Note that this has been a mere outline of string theory (as currently understood). The duality relations allow us to start at any node on the above graph and arrive at any other node. There are more duality relationships which are known, and probably more which are yet to be discovered.

String theory is a vast subject, for which many excellent texts have been written. See especially [Green, Schwarz and Witten, 1987; Polchinski, 1998; Kaku, 1999; Becker, Becker and Schwarz, 2007].

In this chapter we are viewing string theory through the lens of ALE spaces. Thus we will focus on the **IIB** theory, in which the role of D-branes (especially D-strings) is prominent. The above references also deal with D-branes in string theory (and especially in M-theory). However, here we will be mainly following the lead of Johnson in his book *D-Branes* [Johnson, 2003], a follow-up to the paper, "Aspects of type IIB theory on asymptotically locally Euclidean spaces" [Johnson and Myers, 1997].

An ALE space is a smoothed out (disingularized) version of the orbifold $\mathbb{C}^2/\mathcal{m}$, where $\mathcal{m}$ is a finite subgroup of $\mathcal{SU}(2)$. Thus $\mathcal{m}$ is a McKay group and is ADE classified (cf. Chap. 4).

The ADE classification of ALE spaces was conjectured by Michael Atiyah, and was proved by Peter Kronheimer by constructing ALE spaces as hyper-Kähler quotients [Atiyah, 1980; Kronheimer, 1989; 1990].

An ALE space is called a gravitational instanton because it is a solution to the Euclideanized (positive-definite), Ricci-flat version of Einstein's General Relativity field equations:

$$R_{ij} = 0.$$

Stephen Hawking named such spaces gravitational instantons, because they possess self-dual Einstein metrics in analogy with instanton solutions of the self-dual Yang–Mills $\mathcal{SU}(2)$ gauge equations [Hawking, 1977].

An ALE space has many unique properties:

The term ALE means that this 4D space looks like a Euclidean space, except that the boundary at infinity, which is a global property, is not the three-sphere S^3 (the boundary at infinity of $\mathfrak{R}^4$); but rather near infinity the ALE space looks like $S^3/\mathcal{m}$ [Atiyah, 1980; Johnson 2003].

As a hyper-Kähler (H-K) space it has a metric which respects three complex structures, labeled I, J and K, because these complex structures obey the quaternion group formula:

$$I^2 = J^2 = K^2 = IJK = -1.$$

As an H-K space its three Kähler forms $\omega_1, \omega_2, \omega_3$ are closed with respect to the quaternion group and thus provide for three symplectic structures.

As a 4D H-K space, an ALE space is not compact but at infinity looks like $\mathfrak{R}^4/\mathcal{m}$ with boundary $S^3/\mathcal{m}$, in the sense that the singular point becomes desingularized as a "bouquet" of S^2-spheres intersecting each other in the pattern which is dual to the ADE Coxeter graph.

For example, in the E_7 case, this bouquet has the form:

OOOOOO
O

This E_7 bouquet structure is the resolution of the $\mathfrak{R}^4/\mathcal{OD}$ orbifold singularity; and we have been pursuing the thread of the octahedral group since Chap. 2, where we have emphasized the Klein-4 group cosets in the octahedral group as mapping the structure of three fermion families with two quarks and two leptons in each family.

Note especially that the quaternion group Q (which is the double cover of the Klein-4 group) is a normal subgroup of $\mathcal{OD}$, so that three of the six cosets $\mathcal{OD}/Q$ correspond to the three-family structure of the octahedral group. In the $\mathcal{OD}$ case, each of these three families consists of two quarks, two leptons, and their supersymmetry partners. Moreover, the other three cosets of $\mathcal{OD}/Q$ correspond to gauge particles and their supersymmetry partners. In this sense $\mathbb{C}[\mathcal{OD}]$ can be viewed as a 48D supersymmetry algebra.

Thus the three complex structures $\{I, J, K\}$ in $\mathbb{C}^2/\mathcal{OD}$ are uniquely compatible with the three-family structure implicit in the $\mathcal{OD}/\mathcal{Q}$ cosets.

This duality between the three complex structures and the three-family structures encourages us to continue to elaborate the E_7 ALE gravitational instanton.

Accordingly, we will examine the role of the E_7 ALE space, which is the smoothed (desingularized) version of $\mathfrak{R}^4/\mathcal{OD}$ in the type **IIB** string theory. In general, for any ADE version of ALE gravitational instanton, several aspects of the **IIB** theory are uniquely important:

Self-Duality: there is an S (strong-weak coupling) duality which transforms between ordinary type **IIB** strings (which are closed) and D1-branes (which as 1D Dirichlet branes are open). This self-duality can interpolate between a gauge group model on 4D spacetime and conformal field theory on the 2D worldsheet. Such 2D conformal field theories are ADE classified since they invoke the ADE classification of the Kac–Moody Lie algebras, which are infinite dimensional and are classified by the affine ADE graphs [Kaku, 1999].

Note that this **IIB** self-duality is different from the self-dual nature of an ALE space, in which the Riemann metric is self-dual. It is this ALE-metric self-duality that parallels the self-dual Yang–Mills instantons. This accounts for the label "gravitational instanton" for ALE spaces (as mentioned abobe).

Note also that many Yang–Mills instantons are embedded in gravitational instantons. So the relationship is quite close.

Here, however, we are describing S duality, which is a duality between strong and weak interaction strengths. Since strong couplings are very difficult to calculate accurately, it is advantageous to work in the weak coupling form of the theory. The duality guarantees that the physics remains invariant.

Higgs Sector: in parallel to the Higgs field symmetry breaking of electroweak (Yang–Mills) gauge theory, the ALE superpotential breaks supersymmetry. This symmetry breaking (like that of the Higgs field) gives mass to the super-partner particles, and thus makes them extremely difficult to detect. Indeed, such super-partners are a prime candidate for Dark Matter [Binetruy, 2006; Weinberg, 2008].

Note that any ALE superpotential is the germ of an ADE catastrophe. This implies that the zero-set of an ALE superpotential is the ALE orbifold $\mathfrak{R}^4/\boldsymbol{m}$, which can be viewed as $\mathbb{C}^2/\boldsymbol{m}$ imbedded in $\mathbb{C}^3$, whose three parameters $\{X, Y, Z\}$ are the three superpotential (ADE germ) variables.

For $\mathbb{C}^2/\mathcal{OD}$, the E_7 catastrophe germ with the 7D unfolding parameter structure (cf. Chap. 7) is:

$$X^3 + XY^3 + Z^2 + t_1Y + t_2Y^2 + t_3Y^3 + t_4Y^4 + t_5XY + t_6X + t_7.$$

The unfolding parameters are also the moduli of the E_7 gravitational instanton [Anselmi *et al.*, 1994].

Since the gravitational instanton is analogous to the Yang–Mills instanton, it is noteworthy that supersymmetry calculations provide for the negativity of the mass-squared term, which is necessary to derive the Yang–Mills symmetry-breaking Higgs mechanism.

This supersymmetry calculation was also used to accurately predict the mass of the top quark, and the Higgs particle [Kane, 2000].

Clearly, there is a close relationship between Yang–Mills symmetry breaking and supersymmetry breaking. Indeed, superpotential deformations are called "Higgsing" [Okuda and Ookouchi, 2006].

Somewhat more exotic aspects of **IIB** theory are the following:

AdS/CFT duality: the clearest example of this duality is an S (weak-strong) duality between a **IIB** theory (10D) and a conformal field theory (4D).

Thus **IIB** string theory on the 10D space, which is asymptotically the product space of 5D anti-de Sitter space and an internal five-sphere ($\mathrm{Ad}S_5 \otimes S^5$) is S dual to a conformal supersymmetric $\mathcal{SU}(N)$ gauge theory on 4D spacetime, where N is assumed to be very large.

Note that we will revisit this five-sphere in Chap. 9.

Holographic description: by analogy with ordinary holography, in which 3D information is recorded in a 2D space, one can view the 4D conformal boundary of AdS_5 as containing the same information as that of AdS_5, but coded differently.

This holographic description can be generalized to many other similar dualities. Indeed, as a holographic principle, it has been proposed as the basic principle of **M-theory**, which unifies the five known superstring theories.

The AdS/CFT duality was first described by [Maldacena, 1998]. For a detailed description see [Becker, Becker and Schwarz, 2007]. See also [Johnson, 2003].

F-theory: there is an S duality between 12D F-theory and 10D **IIB** superstring theory. F-theory can be seen as a 12D fiber bundle over the 10D **IIB** base, with the 2-torus as fiber. This T^2 fiber pinches off into an ADE-type singularity.

Moreover, the 12D spacetime of F-theory can be compactified on an 8D Calabi–Yau (C-Y) space leaving an uncompacified 4D spacetime. The 8D C-Y space is modeled as the real 8D space $\mathbb{C}^2 \oplus \mathbb{C}^2/m$, where $\mathbb{C}^2/m$ is ADE classified.

This compactification of the 12D F-theory space invokes the corresponding ADE gauge group acting on the 4D spacetime. For an

ADE singularity of rank r, this geometry entails the wrapping of D7-branes in the 8D C-Y space on the r two-spheres of the blown-up singular point. The intersection structure of these 2-spheres form a bouquet which is dual to the ADE Coxeter graph (as pictured for E_7 above).

Note also that the blown-up version of $\mathbb{C}^2/m$ is an ALE space and thus is a gravitational instanton. For this F-theory geometric engineering, see [Heckman, 2010]. See also [Hori *et al.*, 2003].

Given all these special features of the **IIB** superstring theory, it is not surprising that there is a very close relationship between the **IIB** picture and the ADE classification of the ALE spaces.

We will illustrate this relationship in the E_7 case, but analogous relationships hold for all the ADE graphs [Johnson, 1997].

The affine E_7 graph with its balance numbers will play the central organizing role in this development.

1–2–3–4–3–2–1
  |
  2

As we have seen in Chap. 4, these balance numbers are the iireps of the octahedral double group, so that $\mathcal{OD}$ is the finite subgroup of $\mathcal{SU}(2)$ that corresponds to the affine E_7 Lie algebra. As a McKay group, $\mathcal{OD}$ obeys the McKay correspondence:

$$Q \otimes R_i = \Sigma_j A_{ij} R_j.$$

Here Q is the defining 2D (quaternion) representation of the McKay group, as a subgroup of $\mathcal{SU}(2)$ viewed as the set of unit length quaterions in the 4D quaternion algebra.

R_i and R_j are iireps of the McKay group, and A_{ij} is the adjacency matrix of the corresponding ADE affine graph.

This McKay relationship may be read directly from the graph. For example, since the four-node is adjacent to two three-nodes and one two-node, we find that, in accordance with the McKay correspondence formula:

$$2(4) = 3 + 3 + 2 = 8$$

and similarly for all the other nodes.

Another well-known relationship for the affine ADE balance numbers is that the sum of the squares of these numbers is the order of the McKay group. Thus in the $\mathcal{OD}$ case:

$$1 + 4 + 9 + 16 + 9 + 4 + 1 + 4 = 48.$$

This corresponds to the fact that the group algebra of $\mathcal{OD}$ has a matric basis which organizes the group algebra as the direct sum of total matric algebras of dimensions equivalent to these squared balance numbers:

$$\begin{aligned}\mathbb{C}[\mathcal{OD}] = M(1,\mathbb{C}) + M(2,\mathbb{C}) + M(3,\mathbb{C}) + M(4,\mathbb{C}) + M(3,\mathbb{C})\\ +M(2,\mathbb{C}) + M(1,\mathbb{C}) + M(2,\mathbb{C}).\end{aligned}$$

This matric algebra structure was described in Chap. 3 as a duality between the group basis ($\mathcal{OD}$) and the matric basis for these eight matric algebras.

Also, we note that any $\mathrm{M}(n,\mathbb{C})$ algebra consisting of all $(n \times n)$-complex matrices embeds $\mathcal{U}(n)$ as the set of all $(n \times n)$ unitary matrices.

Thus we have embedded the real compact 48D subspace in $\mathbb{C}[\mathrm{OD}]$ as the set of all unitary elements, which is the product of unitary groups:

$$\mathcal{U}(1) \times \mathcal{U}(2) \times \mathcal{U}(3) \times \mathcal{U}(4) \times \mathcal{U}(3) \times \mathcal{U}(2) \times \mathcal{U}(1) \times \mathcal{U}(2).$$

Note that two copies of the standard model gauge group $\mathcal{U}(1) \times \mathcal{SU}(2) \times \mathcal{SU}(3)$ occur as subgroups of this product of unitary groups.

It is significant that type **IIB** superstring theory with the $\mathbb{C}^2/\mathcal{OD}$ gravitational instanton structure should have precisely this set of unitary groups as its gauge group.

Clifford Johnson calls this set of unitary groups the vector multiplet sector of the open-string spectrum of the "D-brane world-volume theory" [Johnson, 1997]. This theory describes D1-branes (Dstrings) sweeping out a 2D worldsheet embedded in the ten spacetime dimensions of the **IIB** superstring theory with $\mathcal{N} = 4$ supersymmetry.

Note that D-strings are open strings which are dual to the "fundamental" closed strings of **IIB** superstring theory which model the gravitational sector of the theory. In general, there are p-dimensional

Dp-branes (Dirichlet branes) which provide boundaries on which open strings can attach.

The vector multiplets that transform in the adjoint representation of the gauge group are bosonic particles analogous to the adjoint representation of the standard model gauge groups, $\mathcal{U}(1), \mathcal{SU}(2), \mathcal{SU}(3)$.

There are also fermionic particles that transform in the fundamental and anti-fundamental representations of the gauge groups.

In the **IIB** case, Johnson calls these "Hypermultiplets II," and these representations can also be read directly from the ADE affine-type graph. These states correspond to the balance numbers and their linkages to each other in the affine graph.

This is because the same adjacency matrix A_{ij} that appears in the McKay correspondence formula appears in the formula for the sum over representations of:

$$\bigoplus_i A_{ij}(n_i, \underline{n}_j).$$

Since the adjacency matrix records the linkage structure of the affine graph, in the $\mathcal{OD}$ case we find the seven links of the extended E_7 graph.

In this application the affine graph should be viewed as a quiver diagram, where the links are dressed with arrows. An arrow-head points to a fundamental representation (n_i), while an arrow-tail corresponds an anti-fundamental representation $(\underline{n}_j)$, which is a charge-parity-time (CPT) reversed copy of the fundamental representation. Accordingly, the affine E_7 quiver-graph can be depicted as:

$$\begin{array}{c} 1\leftarrow 2\leftarrow 3\leftarrow 4\leftarrow 3\leftarrow 2\leftarrow 1 \\ \uparrow \\ 2 \end{array}$$

Thus the matter hypermultiplets (corresponding to these arrows) are

$$(\mathbf{1,2}) + (\mathbf{2,3}) + (\mathbf{3,4}) + (\mathbf{4,3}) + (\mathbf{3,2}) + (\mathbf{2,1}) + (\mathbf{4,2}),$$

where $(\mathbf{a,b})$ is to be read as the representation formed by the tensor product of fundamental representation $\mathbf{a}$ and anti-fundamental representation $\mathbf{b}$.

Note that:

$$(1)(2) + (2)(3) + (3)(4) + (4)(3) + (3)(2) + (2)(1) + (4)(2) = 48$$

which is the order of $\mathcal{OD}$ and matches the dimensionality of the adjoint representation of the gauge group, made up of products of 8 $\mathcal{U}(n)$ groups where n is a balance number. This matching is necessary for supersymmetry.

Since a quiver diagram is any graph with arrows as links, it is important to note that the quivers of finite type are ADE classified [Douglas and Moore, 1996]. In general, the nodes of a quiver correspond to vector-spaces, which are group representation spaces. A finite quiver has only a finite number of inequivalent, irreducible representations (iireps).

This supersymmetry matching works for any of the ADE affine graphs read off in this way. However, we have a special reason for following the thread of the $\mathcal{OD}$ McKay group and its E_7 graph.

This is because the three-family structure of two quarks and two leptons in each of three families is the missing piece in the Standard Model and is also missing from the vast theoretical advances beyond this model provided by string theory. This is sometimes called the "Flavor Problem", because there are two flavors of quarks and leptons in each of the three families (sometimes called generations). The Flavor Problem is regarded by many theoreticians as the deepest problem in particle physics.

As claimed in Chap. 2, the octahedral group provides a template for two quarks and two leptons in each of three families. Moreover, this structure is continued in the octahedral double group.

This three-family structure is provided by three of the six cosets of the Klein-4 group within the octahedral group. In the case of the octahedral double group, the quaternion group (which is the 8-element double cover of the Klein-4 group) provides the cosets for the three families.

Notice especially that the quaternion group is made up of the 8 elements $\{1, -1, I, -I, J, -J, K, -K\}$, which obey the Hamiltonian formula:

$$I^2 = J^2 = K^2 = IJK = -1.$$

And this same quaternion group structure is obeyed by interrelationships of the three complex structures in a H-K manifold such as the ALE smoothed version of $\mathbb{C}^2/\mathcal{OD}$, in which the singularity is blown up into a bouquet of 7 S^2 spheres intersecting each other in the form:

$$\begin{array}{cccccc} O & O & O & O & O & O \\ & & O & & & \end{array}$$

Moreover, the E_7 affine graph determines the projection of 48 D1-branes in $\mathbb{C}^2$ onto the singularity at the origin of $\mathbb{C}^2/\mathcal{OD}$.

When the singularity is blown up into a bouquet of 7 S^2 spheres, the 48 D1-branes map onto the S^2 spheres according to the balance numbers, except that the extra node in the affine graph is now missing.

Thus there are $(n_i)^2$ D1-branes for each balance number n_i in the affine graph, with the missing node indicated:

$$\begin{array}{c} \varnothing - 2 - 3 - 4 - 3 - 2 - 1 \\ | \\ 2 \end{array}$$

Note that there are now 47 D1-branes which wrap the seven two-spheres and are connected to each other by the fundamental strings of the IIB superstring theory [Johnson, 1997].

What happened to the 48th D1-brane at the singularity? It has apparently moved out of the ALE space and into the $\mathfrak{R}^6$ region of the 10D spacetime, $\mathfrak{R}^6 \otimes \mathbb{C}^2/\mathcal{OD}$.

The process of inserting 48 D1-branes into the projection $\mathbb{C}^2 \to \mathbb{C}^2/\mathcal{OD}$ can be repeated, so that presumably any number of D1-branes can be pumped into $\mathfrak{R}^6$ via the blow-up of the singularity in $\mathbb{C}^2/\mathcal{OD}$ into 7 S^2-spheres, as described above.

This is but a small part of what is called "geometric engineering" [Johnson, 2003; Hori *et al.*, 2003]. In this much more involved picture, D3-branes and D7-branes (and others) play a significant role.

In summary we can list the various ADE classified structures that have been invoked in this chapter. This is a good demonstration of the intertwining of these ADE structures, and justifies the claim that we are engaged in a vast unification called ADEX theory.

(1) McKay correspondence groups — $\mathcal{OD} \leftrightarrow$ affine E_7
(2) ALE spaces (gravitational instantons) — desingularized $\mathbb{C}^2/\mathcal{OD}$
(3) Coxeter graph dual $\leftrightarrow$ bouquet of two-spheres
(4) McKay group algebra $\mathbb{C}[\mathcal{OD}] \leftrightarrow$ matric basis $\leftrightarrow 8\,\mathcal{U}(n)$s
(5) McKay group iireps $\leftrightarrow$ **IIB** hypermultiplets
(6) Catastrophe germs $\leftrightarrow$ superpotential $\leftrightarrow$ Higgsing SUSY
(7) 2D conformal field theories $\leftrightarrow$ D-string worldsheet fields
(8) Kac–Moody Lie algebras $\leftrightarrow$ affine ADE Coxeter graphs
(9) Finite quivers $\leftrightarrow$ **IIB** hypermultiplets
(10) Knots and links $\leftrightarrow$ $\mathrm{AdS}_5 \otimes S^5$.

This last item was only hinted at in this chapter, and it will be described in Chap. 9.

Chapter 9

Knots and Links and Braids

Knot theory is a vast enterprise in mathematics, with many recent and surprising discoveries making intimate connections between mathematics and physics, especially the physics of low-dimensional conformal field theory. See especially the work of Witten relating the knot invariant called the Jones polynomial with conformal field theory on 3D space [Atiyah, 1990; Kauffman, 1991].

Here, however, we will be describing a lesser known aspect of knot theory: the ADE classification of those knots (and links) which are derivable from the ADE singularities as described in Chap. 8.

Given the McKay group m, the ADE knots can be described as substructures in the intersection of $\mathbb{C}^2/m$ with S^5 a sufficiently small sphere around the origin of $\mathbb{C}^3$ in which $\mathbb{C}^2/m$ is embedded.

We will concentrate especially on the E_7 case, and its relationship with **IIB** superstring theory. Since the E_7 singularity structure corresponds to the action on $\mathbb{C}^2/\mathcal{OD}$, we will begin with a description of $\mathcal{OD}$ as a subgroup of $\mathcal{SU}(2)$, which geometrically is S^3 as the unit sphere in $\mathfrak{R}^4 (= \mathbb{C}^2)$.

The octahedral double group, $\mathcal{OD}$, is a 48-element group with the 24-element tetrahedral double group, $\mathcal{TD}$, as a normal subgroup. Now $\mathcal{TD}$ is especially easy to describe as a set of 24 points in S^3.

Coxeter calls this structure the 24-cell and points out that as a regular polytope in 4D Euclidean space, it is special since it has no analogue in either higher or lower dimensional spaces [Coxeter, 1973]. The 24-cell is self-dual in the sense that it has 24 vertices and 24 3D facets (i.e. cells). Each of these 24 cells is a regular octahedron. In this sense it is analogous to the tetrahedron, which has four vertices and four (triangular) faces, so that the tetrahedron is self-dual: the face center-points are the vertices of another tetrahedron.

We note in passing that the 24 vertices of the 24-cell are also the 24 roots of the D_4 Lie algebra, and thus correspond to the optimal lattice packing of 24 unit spheres around a central sphere in 4D Euclidean space [Conway and Sloane, 1988].

The description of the 48 vertices of the $\mathcal{OD}$ polytope in $\mathfrak{R}^4$ is more complicated, because it is not regular. It consists of two 24-cells placed reciprocally to each other, so that a structure is created with 48 vertices and 288 tetrahedral 3D faces. The dual polytope to this structure is a polytope with 288 vertices and 48 3D faces, each of which is a truncated cube [Coxeter, 1973; Duval, 1964].

Note that both 48 and 288 can be derived from the E_7 balance numbers on the affine E_7 Coxeter graph:

```
1–2–3–4–3–2–1
      |
      2
```

where 48 is the sum of the squares of these balance numbers, while 288 is the product of these balance numbers. These balance numbers are also the iireps of the $\mathcal{OD}$ group (cf. Chap. 4, "The McKay correspondence").

In order to elucidate the structure of $S^3/\mathcal{OD}$, we will concentrate on the structure of the truncated cube, 48 of which are the 3D faces of the polytope dual to the 48 vertex polytope, which locates the 48 elements of the $\mathcal{OD}$ group in $\mathcal{SU}(2)$, i.e. in S^3.

In 3D Euclidean space, the cube and the octahedron are dual to each other: the cube has six faces and eight vertices, while the octahedron has eight faces and six vertices. Thus inside a cube one can inscribe an octahedron by joining up the central points of the six faces of the cube. Likewise inside an octahedron a cube can be inscribed by joining up the central points of the eight faces of the octahedron. The symmetry group of the octahedron (and the cube) is the octahedral group.

Now one can truncate the cube by setting it within an octahedron in such a way that the eight corners of the cube are cut off by the octahedron, so that these eight corners are replaced by equilateral triangles. This transforms each of the six cube faces into perfect octagons. The truncated cube has 24 vertices, 14 faces (six triangles, and eight octagons), and 36 edges. (See "Truncated Cube" in Wikipedia.)

In order to elucidate the topology of the singularity structure of $\mathbb{C}^2/\mathcal{OD}$, we find it concentrated in the boundary at infinity, which is $S^3/\mathcal{OD}$. According to [Duval, 1964], this topology is that of the truncated cube with its faces identified with certain twists.

To understand this structure, it is useful to think of analogous but much simpler structures. For example, a 2D torus can be constructed by identifying the opposite edges of a square (as in the figure below). First by identifying the edges A and A' with each other, and not identifying the other two edges, we would have a cylinder: $S^1 \otimes \mathfrak{R}$. But now if we also identify the right and left edges with each other we will have a 2D torus: $T^2 = S^1 \otimes S^1$.

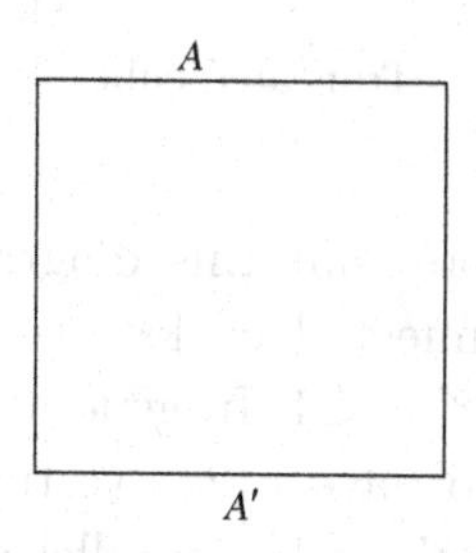

But if we reverse the orientation of the A edges with respect to each other when we identify these edges, ($A \rightarrow$) with ($\leftarrow A'$), we will have the topology of a Klein bottle [Weeks, 1985].

Likewise, a cube with opposite faces identified has the topology of a 3D torus: $T^3 = S^1 \otimes S^1 \otimes S^1$.

Of course, the truncated cube identifications of its faces are more complicated. According to Duval the opposite octangular faces must be identified while undergoing a rotation of 45°; and the opposite triangular faces must be identified while undergoing a rotation of 120°.

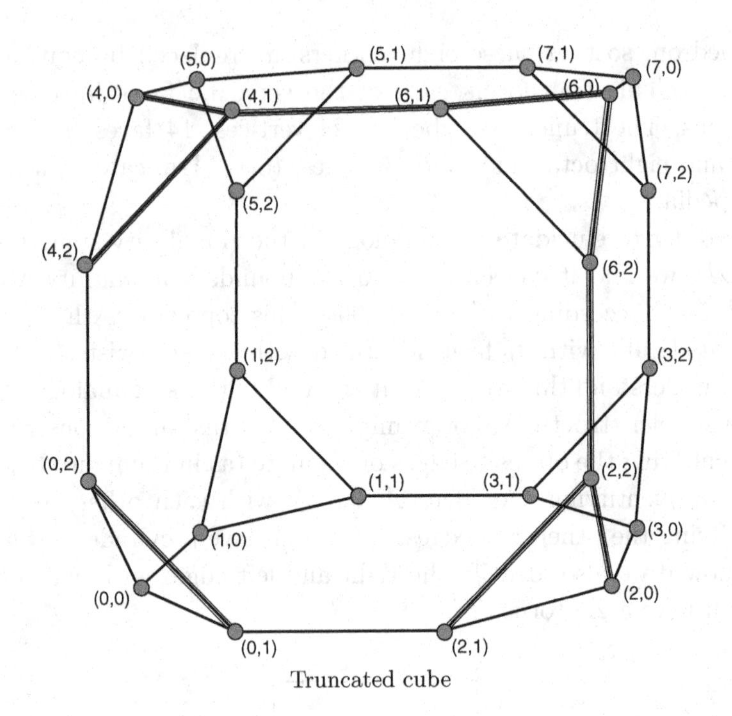

Truncated cube

It is interesting to note that this diagram is the Cayley graph of the group of cube-connected cycles of order three, so that the number of vertices is $3(2^3) = 24$. In general such a cube-connected-cycle Cayley graph would have $n(2^n)$ vertices. Such Cayley graphs describe the network topology for parallel processing of computers [Preparate and Vuillemin, 1981].

Here we indulge in a brief digression on the 24-element octahedral group, $O = \mathcal{OD}/\{\pm 1\}$.

Note that the truncated cube Cayley graph does not correspond to the octahedral group, Moreover, the polytope for the octahedral group Cayley graph is not the dual of the truncated cube, but is nevertheless quite analogous. Just as one uses the octahedron to truncate the cube, one can use the cube to truncate the octahedron. And indeed, the Cayley graph for the octahedral group is the truncated octahedron. In this case there are also 14 faces: six squares and eight hexagons. In fact, the six square faces correspond to the six cosets of the Klein-4 group within the octahedral group, as described in

Chap. 2. (Since Klein-4 is a normal subgroup of O, the right and left cosets are the same.) The claim made in Chap. 2 is that the three odd cosets of the Klein-4 group correspond to the three fermion families with two quarks and two leptons in each family. So it is rather nice that this coset structure shows up so prominently in the octahedral group Cayley graph as the six square faces of the truncated octahedron.

After this digression on the octahedral group, we resume the treatment of the octahedral double group ($\mathcal{OD}$), which is the affine E_7 McKay group.

The truncated cube corresponds to the structure of $S^3/\mathcal{OD}$ in the sense that the twisted identification of the truncated-cube's faces provides the structure of $S^3/\mathcal{OD}$, especially revealing its complicated topology. We can think of this topology as that of a kind of 3D hyperknot (or hyperlink) in S^5, just as an ordinary knot is a 1D manifold in S^3. Then, since S^5 is the unit sphere in $\mathfrak{R}^6(=\mathbb{C}^3)$, we can write the following equation, cf. [Duval, 1964; Milnor, 1968; Durfee, 1975]:

$$K^3 = S^3/\mathcal{OD} = \mathbb{C}^2/\mathcal{OD} \cap S^5,$$

where K^3 is the truncated cube hyperknot, which is equivalent to the intersection of S^5 with the hyper-Kähler orbifold $\mathbb{C}^2/\mathcal{OD}$. We are assuming that S^5 is a small unit sphere centered on the origin of $\mathbb{C}^3$, remembering that $\mathbb{C}^2/\mathcal{OD}$ is embedded as the zero set of the E_7 polynomial, which is the germ of the E_7 catastrophe polynomial:

$$X^3 + XY^3 + Z^2,$$

where X, Y and Z, as complex parameters, are the bases of $\mathbb{C}^3$, so that the zero set of this E_7 germ is naturally embedded in $\mathbb{C}^3$.

In order to describe the topology of $S^3/\mathcal{OD}$, we need to reveal the structure of the knot (or link) embedded in the truncated-cube hyperknot K^3. In this case, we need to look at the intersection of the zero set with unit-sphere S^3 in $\mathfrak{R}^4(=\mathbb{C}^2)$, The zero set of X^3+XY^3 has, as a branched covering, the zero set of the full E_7 catastrophe germ, $X^3+XY^3+Z^2$ [Rolfsen, 1976; Barth, Peters and Van de Ven, 1984].

Here we assume that S^3 is a very small sphere centered around the origin of $\mathbb{C}^2$ where the singularity of the $(X^3 + XY^3)$-zero-set is located, and that the intersection with this very small S^3 will reveal a knot structure [Duval, 1964].

However, we can factor this polynomial in complex X and Y into:

$$(X)(X^2 + Y^3).$$

Thus we expect the knot structure to be a link in the form of two knots linked together. We see immediately that the $(X^2 + Y^3)$-factor will correspond to a torus knot of type (2,3), which is a trefoil knot.

This means that the trefoil knot is equivalent to a closed curve winding around the torus 2-times in one direction, while in the transverse direction it winds around the torus 3-times.

The other factor (X) must be the unknot, or simply a circle linking through the structure of the (2,3)-trefoil knot. In the topology of this structure, the linking is accomplished by a so-called Dehn filling (or Dehn surgery), which puts a solid torus inside the trefoil knot, so that the unknot-circle can be inserted to link with the trefoil on the boundary of the solid torus [Rolfsen, 1976].

According to Rolfsen (p. 272) Dehn surgery on torus knots has been completely classified by [Moser, 1971]. There are three types:

Lens spaces, double Lens spaces, and Seifert manifolds. This must be the rather ubiquitous ADE-classification.

So, drawing on [Milnor, 1968] and [Orlik, 1972], we can make the following classifications:

(1) Lens spaces: S^3/C_{k+1} cyclic-group; which are diffeomorphic to the intersection of the zero-set of the complex polynomial $X^{k+1} + Y^2 + Z^2$. This is the A_k type McKay group case (cf. Chap. 7).
(2) Double lens spaces: S^3/Q_{k-2} quaternion type group (binary dihedral group); which is diffeomorphic to the intersection of S^5 with the zero-set of $X^2Y + Y^{k+1} + Z^2$. This is the D_k type McKay group case.
(3) Seifert manifolds of types $\mathcal{TD}, \mathcal{OD}$ and $\mathcal{ID}$:

($\mathcal{TD}$) Tetrahedral double space: $S^3/\mathcal{TD}$ (binary tetrahedral group), which is diffeomorphic to $\{X^3 + Y^4 + Z^2 = 0\} \cap S^5$.

The knot in $S^3/\mathcal{TD}$ is the (2,3)-torus knot, a trefoil. This is the E_6 type McKay group case.

($\mathcal{OD}$) Octahedral double space: $S^3/\mathcal{OD}$ (binary octahedral group), which is diffeomorphic to $\{X^3 + XY^3 + Z^2 = 0\} \cap S^5$. The knot in $S^3/\mathcal{OD}$ is the link of the (2,3)-trefoil with an S^1 (unknot), as described above. This is the E_7 type McKay group case.

($\mathcal{ID}$) Icosahedral double space: $S^3/\mathcal{ID}$ (binary icosahedral group), which is diffeomorphic to $\{X^3 + Y^5 + Z^2 = 0\} \cap S^5$. The knot in $S^3/\mathcal{ID}$ is the (2,3)-trefoil knot. This is the E_8 type McKay group case.

Thus we see that among these E-type topologies, E_7 is special in the sense that E_7 has a link-type topology. This is related to the structure of the McKay group itself, which determines whether or not the topology of the knot is derived from a polynomial of the form:

$$X^p + Y^q + Z^r,$$

where p, q, r are integers. The zero-set of this polynomial is a 2D complex space called a Brieskorn variety, $V(p, q, r)$. The intersection $V(p, q, r) \cap S^5$ is the smooth, compact 3D manifold called the Brieskorn manifold $M(p, q.r)$ [Milnor, 1968, 1975; Kauffman, 1987].

The Brieskorn manifold $M(2, 3, r)$ is homeomorphic to the r-fold branched covering of S^3, branched along a torus knot of type (2,3).

The existence of $M(2, 3, r)$ is based on the structure of the McKay group. That is, for every finite subgroup $\mathcal{m}$ of $\mathcal{SU}(2)$, with the commutator subgroup

$$\Pi = [\mathcal{m}, \mathcal{m}],$$

where for elements g and h in $\mathcal{m}$, $[g, h] = g^{-1}h^{-1}gh$.

$\mathbb{C}^2/\Pi$ is mapped into a Brieskorn variety, $V(p, q, r)$, so that S^3/Π is diffeomorphic to the Brieskorn manifold $M(p, q, r)$.

Other than the trivial A_k case of the cyclic groups (which are all commutative), we have three interesting cases:

$$Q_4 = [\mathcal{TD}, \mathcal{TD}];\ Q_4 \text{ is the } D_4 \text{ McKay group, so that:}$$
$$\mathbb{C}^2/Q_4 = V(2, 3, 3) \quad \text{and } S^3/\Pi = M(2, 3, 3).$$

$$\mathcal{TD} = [\mathcal{OD}, \mathcal{OD}]; \mathcal{TD} \text{ is the } \mathrm{E}_6 \text{ McKay group, so that:}$$
$$\mathbb{C}^2/\mathcal{TD} = V(2,3,4) \quad \text{and } S^3/\mathcal{TD} = M(2,3,4).$$

$$\mathcal{ID} = [\mathcal{ID}, \mathcal{ID}]; \mathcal{ID} \text{ is the } \mathrm{E}_8 \text{ McKay group, so that:}$$
$$\mathbb{C}^2/\mathcal{ID} = V(2,3,5) \quad \text{and } S^3/\mathcal{ID} = M(2,3,5).$$

Thus although $\mathcal{OD}$ provides $\mathcal{TD}$ as a commutator subgroup, $\mathcal{OD}$ itself is not the commutator subgroup of any other group, $\mathcal{ID}$ being its own commutator subgroup. So clearly, $\mathcal{OD}$ is the odd man out here.

We note in passing that here we make contact with the beginnings of group theory in Galois' proof of the unsolvabillity (by radicals) of the quintic equation. His proof depends on the fact that the alternating-5 group (the Icosahedral group) has no normal subgroup (other than itself and the identity element). The octahedral group (the symmetric-4 group) has the chain of normal subgroups:

$$\mathrm{O} \to T \to K4 \to \{e\}$$

so that O is a solvable group, whereas I (and any alternating-n group, where $n \geq 5$) is not solvable, and reflects the general unsolvability (by radicals) of equations of degree higher than 4 [Birkhoff and MacLane, 1965].

The ubiquity of the (2,3)-trefoil knot is remarkable: Q_4, $\mathcal{TD}$ and $\mathcal{ID}$ all correspond to the trefoil topology in $\mathbb{C}^2/\Pi$. Even $\mathbb{C}^2/\mathcal{OD}$ has the topology of the (2,3)-trefoil linked with S^1.

Since Brieskorn varieties $V(p, q, r, \ldots, x_n)$ can be defined for higher dimensional polynomials, they are a doorway into the study of higher dimensional hyperknot topologies [Kauffman, 1987]. However, these higher-dimensional Brieskorn varieties go beyond the ADE-type structures we are describing here.

Rather, we will go on to explore the remarkable ADE classification of Braid groups corresponding to the ADE classification of singularities, which in turn corresponds to the ADE classification of knots. Of course, these classifications are a very special substructure within the set of all these objects.

Let us, however, set out a preliminary diagram indicating the relationships between various ADE classifications, which we will be describing.

$$\begin{array}{ccccc} \text{Knots} & \leftrightarrow & \text{Braids} & \leftrightarrow & \pi_1(\mathfrak{h}^{\mathfrak{reg}}/W) \\ \downarrow\uparrow & & \downarrow\uparrow & & \downarrow\uparrow \\ \mathbb{C}^2/m & \leftrightarrow & \text{Bouquet of } (S^2)s & \leftrightarrow & \text{Coxeter graphs} \end{array}$$

Here Knots includes Links; and $\mathfrak{h}^{\mathfrak{reg}}/W$ is the space of regular orbits of the Coxeter group W, so that $\pi_1(\mathfrak{h}^{\mathfrak{reg}}/W)$ is the fundamental group of this space.

Note that there are three groups whose interrelations are implicit in this diagram:

(1) m is the McKay group as a finite subgroup of $\mathcal{SU}(2)$.
(2) W is the Coxeter group corresponding to the ADE graph.
(3) $\pi_1(\mathfrak{h}^{\mathfrak{reg}}/W)$ is the corresponding Braid group [Brav and Thomas, 2011].

There is also another relationship. Since $S^3/m(= \mathcal{SU}(2)/m)$ is embedded in $\mathbb{C}^2$, it is worth pointing out that m is the fundamental group of S^3/m, which reveals the hyperknot topology as described above [Duval, 1964].

These observations require some elaboration:

There is a well-known relationship between knots (and links) and braids. This is because any knot or link can be obtained from a braid, simply by joining the ends of the braids. This was proved by J.W. Alexander in 1923 [Kauffman, 1991]. Note, however, that different braids can generate the same knot (or link).

For example, consider the Symmetric group S_n as the set of all permutations of n things. For n strands, numbered $\{1, \ldots, n\}$, each element of S_n reorders these numbers. Thus we can see that each element of S_n records the beginning and ending of a braiding of n strands, assuming that we always braid over the strands as we braid the n strands (say from left to right).

This S_n group would be a homomorphic image of B_n, the braid group on n strands, which takes into account the freedom to braid over or under at each move through the n strands. Thus B_n would have an infinite number of elements, rather than the $n!$ elements of S_n.

Since S_n is the Coxeter group of type A_{n-1}, we might wonder whether there is a homomorphism from the D and E type Coxeter groups to Braid groups.

This question is answered by Artin in his generalization of the Braid group on n strands B_n.

The A-type Coxeter group S_n is the homomorphic image of B_n according to the short exact sequence:

$$1 \to P_n \to B_n \to S_n \to 1,$$

where P_n is the pure braid group on n strands; and the exact sequence of groups implies that S_n is the quotient of B_n/P_n.

Artin's generalization of the braid group B_n has the analogous short exact sequence:

$$1 \to P \to B \to W \to 1,$$

where W is any Coxeter group, so that the A-type Coxeter group (S_n) is the special case where B is the braid group on n-strands B_n.

Here we are considering only the ADE type Coxeter groups. In these cases we can use the adjacency relations of the ADE Coxeter graphs to provide a presentation of the (generalized) Braid groups of Artin, which we will call B_W.

This is because the presentations of these two groups W and B_W are very similar.

(1) ADE Coxeter group W presentation:

For a Coxeter graph $G(W)$ of rank n, the Coxeter group W has n generators g_i, where i runs from 1 through n. The relations between these generators are:

$$(g_i)^2 = 1 \text{ (each generator is a reflection);}$$

$$g_i g_j = g_j g_i \text{ (if the nodes } i, j \text{ are not adjacent in } G(W)),$$

$$g_i g_j g_i = g_j g_i g_j \text{ (if the nodes } i, j \text{ are adjacent in } G(W),$$

(2) ADE Artin group B_W presentation:

For a Coxeter graph $G(W)$ of rank n, the Artin group B_W has n generators b_i corresponding to the nodes of the Coxeter graph. Since B_W is not a reflection group, the reflection relations are omitted. The remaining relations are exactly analogous to those of the Coxeter group W.

$$b_i b_j = b_j b_i \text{ (if the nodes } i, j \text{ are not adjacent in } G(W));$$
$$b_i b_j b_i = b_j b_i b_j \text{ (if the nodes } i, j \text{ are adjacent in } G(W)$$

[Brav and Thomas, 2011; Allcock, 2002; Van der Lek, 1983; Coxeter and Moser, 1965].

In this B_W presentation, the adjacent node rule is interpreted as a braiding. This braiding takes place via paths in $\mathbb{C}^n$, the complex Cartan subalgebra space $\mathfrak{h}$ where the reflection hyperplanes reside. These n basic hyperplanes generate the Coxeter reflection group W. The reflections in the basic hyperplanes create the Coxeter arrangement X of $kn/2$ hyperplanes (where k is the Coxeter number, and n the rank of the Coxeter graph $G(W)$).

The pure Artin group P_W is the fundamental group of the complement $\mathfrak{h}_o = \mathfrak{h} - X$ of the reflection hyperplanes X generated by W. Thus we can write: $P_W = \pi_1(\mathfrak{h} - X)$.

Also the Artin group B_W is the fundamental group of the quotient, $\mathfrak{h}_0/W$. Thus $B_W = \pi_1(\mathfrak{h}^{\mathfrak{reg}}/W)$, since the regular orbits of W in $\mathfrak{h}$ are transformed by the action of W into the regular points of $\mathfrak{h}/W$, which are not in the discriminant hypersurface Δ. This hypersurface is the image generated by the action of W on the hyperplanes X in $\mathfrak{h}$.

This action generates discrete orbits, which become points of the orbifold $\mathfrak{h}/W$. Thus Δ consists of the special orbits of W, whose cardinality is less than the order of W. (The regular orbits of W consist of points outside the hyperplane complex X, and have cardinality equal to order of W.) This corresponds to the ADE catastrophe structures described in Chap. 7, where we call the discriminant Δ the separatrix Σ.

Since $\pi_1(\mathfrak{h}^{\mathfrak{reg}}/W)$ is the first homotopy group of the manifold $\mathfrak{h}^{\mathfrak{reg}}/W$, we can think of the homotopic paths (which correspond to regular orbits in $\mathfrak{h}$) as braidings [Allcock, 2002].

Thus the Coxeter group W corresponding to the Coxeter graph $G(W)$ is a homomorphic image of the Artin group corresponding to the Coxeter graph $G(W)$. That is, we have the short exact sequence:

$$1 \to P_w \to B_w \to W \to 1$$

so that:

$$B_w/P_w = W.$$

An alternative method of generating B_w from the ADE Coxeter graph $G(W)$ is via the dual to this graph, which is a bouquet of 2-spheres. Each node $G(W)$ is replaced with the sphere S^2, which can be considered as the complex projective line $P^1(\mathbb{C})$. The adjacency relations of $G(W)$ then become intersections at a point between the 2-spheres.

For example in the E$_7$ case, the dual to the E$_7$ Coxeter graph is made up a bouquet of 2-spheres with the structure:

OO8OOO dual to o–o–o–o–o–o
(with one further O/o attached below the third node in each)

It is the spherical twists t_i (which are generalized Dehn twists) on the 2-spheres S_i that generate the braiding group B_w. There are two rules analogous to the rules for the Coxeter graph $G(W)$ described above [Brav and Thomas, 2011].

(1) If S_i and S_j are not adjacent in the W-bouquet of 2-spheres, then $t_i S_j \approx S_j$.

(2) If S_i and S_j are adjacent in the W-bouquet of 2-spheres, then $t_i t_j S_i \approx S_j$.

Moreover, there is an intimate relationship between the ADE bouquet-of-sphere structures and the ALE structures derived from $\mathbb{C}^2/m$, where m is the ADE McKay group (as described in Chap. 8). This is because $\mathbb{C}^2/m$, is a variety with an ADE-type singularity at the origin of $\mathbb{C}^3$ in which $\mathbb{C}^2/m$, is embedded. The ALE (Asymptotically Locally Euclidean) manifold of ADE-type is generated by the resolution of the singularity in $\mathbb{C}^2/m$. In this resolution, the singular point blows up into a bouquet of 2-spheres having a structure dual to that of the ADE Coxeter graph, as in the E$_7$ example described above (cf. Chap. 8).

In this way we implicate the ADE-type knot (and link) structures described above. This is because the intersection of $\mathbb{C}^2/m$ with S^5

the origin of $\mathbb{C}^3$ reveals the topology of the ADE-type hyperknot. Furthermore, embedded in this hyperknot is an ordinary knot as the intersection of the hyperknot with S^3 in $\mathbb{C}^2$.

The question now arises: How does the braiding described by the Artin group B_w relate to the braiding of the ADE knots? It would appear that the ADE knots correspond to the ordinary braid group B_n, which is the special case of B_w. Here W is the A-type Coxeter group S_n, which is also the homomorphic image of B_n, the braid group on n strands.

Given the intimate relationships between the various ADE structures depicted in the diagram above, it must be that the knot groups of A-type are embedded in the Artin groups of D and E type.

For example, in the E_7 case we have:

A_6 o—o—o—o—o—o as a subgraph of

E_7 o—o—o—o—o—o (with one further node o attached below the third node)

so that the A_6 Coxeter group S_7 is a subgroup of the E_7 Coxeter group of order: (product of E_7 balance numbers)(7!)(2). Here 7! is the order of S_7, which permutes the seven basic reflection hyperplanes that generate the $\mathcal{E}_7$ Coxeter group. Moreover, we have:

$$S_7 \subset B_7 (= B_{W(\mathrm{A6})}) \subset B_{W(\mathrm{E7})}.$$

Thus we expect that the E_7 braidings by way of the A_6 braidings will generate the E_7 type knot, which is the link consisting of the (2,3) torus knot (the trefoil) linked with a circle (an unknot) threaded through the trefoil.

In general, there is a mapping from any Artin group of ADE-type to the ADE-type knot corresponding to the ADE-type singularity. It is the blow up of this singularity into an ADE-type bouquet of 2-spheres that provides this mapping. Indeed the generalized Dehn twists on these 2-spheres provide a braiding which generates the ADE-type knot. The fact that different braidings can generate the same knot is an important consideration in this mapping picture.

It is especially noteworthy that the geometric and algebraic structures needed to provide these mappings are to be found in the complex Lie algebras of ADE-type (cf. Chap. 5). Thus the key bridge between all these mappings is the set of ADE Coxeter graphs.

To summarize, we have the intimate mapping relationships between the various ADE-type structures in the diagram, which we present again here, with the addition of the Lie algebra and Coxeter group of ADE type:

$$\begin{array}{ccccccc} \text{Knot} & \leftrightarrow & \text{Braid } (B_w) & \leftrightarrow & \pi_1(\mathfrak{h}^{\mathfrak{reg}}/W) & \leftrightarrow & \text{Lie Algebra (ADE)} \\ \downarrow\uparrow & & \downarrow\uparrow & & \downarrow\uparrow & & \downarrow\uparrow \\ \mathbb{C}^2/m & \leftrightarrow & \text{Bouquet of } (S^2)\text{s} & \leftrightarrow & \text{Coxeter graph} & \leftrightarrow & \text{Coxeter group } W \end{array}$$

The set of intimate relationships described by these ADE graphs is provided in detail by the following papers: [Siedel, 1999; Siedel and Smith, 2006; Allcock, 2002; Brav and Thomas, 2011].

Chapter 10

Twistors and ALE Spaces

Twistor theory originated in 1967 with the work of Roger Penrose, who intended it as an approach to quantizing gravity. Earlier work had already introduced a spinor formulation of Einstein's theory of general relativity as a bridge to the quantization of gravity [Weinberg, 1972]. The much more radical idea of twistors entailed the replacement of 4D spacetime with a 4D complex twistor space as the more basic structure, from which curved 4D spacetime would emerge. There is a vast literature, but see especially: [Hughston and Ward, 1979; Penrose and Rindler, 1984, 1986; Ward and Wells, 1990; Penrose, 2005]. The later work is copiously illustrated with drawings by Penrose himself, and is a very useful adjunct to the more algebraic structures necessary for an exact representation.

In an abstract sense, a twistor is an element of $\mathbb{C}^4$, with signature (2,2), whose Hermitian norm is invariant under the action of the Lie group $\mathcal{SU}(2,2)$, which is the 2-fold cover of the conformal group $\mathcal{SO}(4,2)$. This is analogous to the description of a spinor as an element of $\mathbb{C}^2$ acted on by the Lie group $\mathcal{SU}(2)$, whose generators are the three Pauli spin matrices, which are the three basis elements of the Lie algebra $\mathfrak{su}(2) = \mathfrak{sl}(2\mathbb{C})$. In this sense a twistor is a higher dimensional spinor.

The ordinary rotation group in $\mathfrak{R}^3$ is the $\mathcal{SO}(3)$ group. However, $\mathcal{SU}(2)$ is the 2-fold cover of $\mathcal{SO}(3)$. That is $\mathcal{SO}(3) = \mathcal{SU}(2)/\{\pm 1\}$, so that a 360° rotation in $\mathbb{C}^2$ corresponds to a 720° rotation in $\mathfrak{R}^3$ [Gilmore, 1974]. This relationship accounts for the spin-$\frac{1}{2}$ nature of the fermions (such as protons and electrons). The spin-1 nature of the photon is based on the 3D adjoint representation of $\mathcal{SU}(2)$. And higher spin states are based on higher dimensional representations of $\mathcal{SU}(2)$.

Furthermore, the Lie groups $\mathcal{SU}(2)$ and $\mathcal{SL}(2, \mathbb{C})$ have the same Lie algebra; and we note that $\mathcal{SL}(2, \mathbb{C})$ is the 2-fold cover of the Lorentz group $\mathcal{SO}(1, 3)$. So we have the parallelism:

$$\mathcal{SO}(1,3) = \mathcal{SL}(2, C)/\{\pm 1\} \text{ for spinors in } \mathbb{C}^2$$
$$\mathcal{SO}(2,4) = \mathcal{SU}(2,2)/\{\pm 1\} \text{ for twistors in } \mathbb{C}^4.$$

There is a sense in which $\mathcal{SU}(2)$ remains the key link between these structures. This is because Penrose introduced the word "twistor" to describe the twisting Clifford parallel structure of geodesic curves living on the S^3 manifold of $\mathcal{SU}(2)$. Note that $\mathcal{SU}(2)$ is the unit sphere S^3 in Hamilton's quaternion algebra $\mathfrak{H}(= \mathbb{C}^2 = \mathfrak{R}^4)$ [Penrose, 1978, 2005].

These geodesic curves are the flow lines of a left-invariant vector field living on $\mathcal{SU}(2)$. The set of all left-invariant vector fields on a Lie group manifold constitutes the Lie algebra of the Lie group. Thus the original "twistor" is the closest thing we have to a picture of one of the three basis elements of $\mathcal{SU}(2)$, the Lie algebra of $\mathcal{SU}(2)$.

This image is a conformal mapping of $\mathcal{SU}(2)$ as a 3-sphere to the Euclidean space $\mathfrak{R}^3$ (plus the point at infinity). This is analogous to the conformal ("stereographic") mapping of S^2 to $\mathfrak{R}^2$ (plus the point at infinity). Using polar coordinates in $\mathfrak{R}^2$, we can map parallels in S^2 to concentric circles in $\mathfrak{R}^2$, while meridians in S^2 are mapped to straight lines raying out from the origin of $\mathfrak{R}^2$ (which corresponds to the "south pole" of S^2). The "north pole" of S^2 is mapped to the point at infinity. Note that if we replace $\mathfrak{R}^2$ with $\mathbb{C}^1$, then S^2 is called the Riemann sphere, and this Riemann sphere will play the role of the celestial sphere in twistor theory.

The celestial sphere as S^2 is implicit in ordinary 4D Minkowski space, where the light-cone of null lines is a double cone made up of a succession of ever larger copies of S^2 emanating out in both positive and negative time directions from the point at the origin. Note that this is the full 4D version of the (ordinarily depicted) 3D projection of Minkowski space, which contains the usual light-cone made up of successively enlarging copies of S^1 circles.

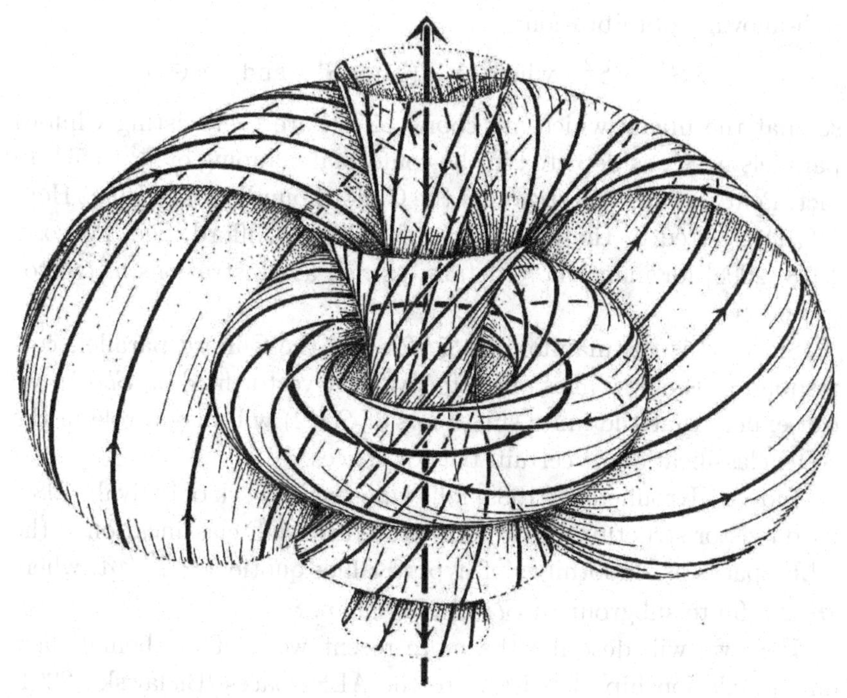

Twistor (figure reprinted with permission from R Penrose).

Thus an observer at the origin of Minkowski space is always at the origin of a light-cone, so that the observer's celestial sphere is a copy of S^2 in his past light-cone.

The Riemann sphere S^2, regarded as a conformal mapping of $\mathbb{C}$, can be mapped directly to projective twistor space $\mathcal{PT}$, which is a complex 3D space as a projection from the full twistor space $\mathcal{T}$ (a complex 4D space). Since the light-cone in Minkowski space is made up of null rays, the Riemann sphere must be mapped to a copy of S^2 in $\mathcal{PN}$, the projective null subspace of the projective twistor space $\mathcal{PT}$. However, the slice of $\mathfrak{R}^3$ representing a fixed time ("now") in Minkowski space must be represented as S^3 in $\mathcal{PN}$, so that the null lines in Minkowski space become a Robinson congruence of Clifford parallels in S^3, as described above. The relationship between the Riemann sphere S^2 and the twistor sphere S^3 is described by the well-known Hopf fibration:

$$\pi\colon S^3 \to S^2 \quad \text{where} \quad \pi^{-1}(x) = S^1 \quad \text{and} \quad x \in S^2$$

so that the fibers, which are copies of S^1, are the twisting Clifford parallels in S^3, as depicted in the conformal mapping of S^3 to $\mathfrak{R}^3$. In fact, Penrose calls this structure the Clifford bundle, since Heinz Hopf (1931) referred to the prior work of William Clifford. See [Penrose, 1978, 2005] for many more details, as well as illustrations of twistor geometry.

Since S^3 is the manifold of $\mathcal{SU}(2)$, and the Clifford parallels correspond to the flow lines of a left-invariant vector field on $\mathcal{SU}(2)$, we can expect that the finite subgroups of $\mathcal{SU}(2)$ will play a role in the ADE classification of certain twistor spaces.

Indeed, [Kronheimer, 1989] following up the work of [Atiyah, 1980] used twistor structures in order to prove the ADE classification of the ALE spaces as smoothings of hyper-Kähler quotients, $\mathbb{C}^2/\mathcal{m}$, where $\mathcal{m}$ is a finite subgroup of $\mathcal{SU}(2)$. Cf. Chap. 8.

Here we will describe the more recent work of mathematicians on the relationship of twistors to the ALE spaces [Bielawski, 2001; Dunajski and Mason, 2003]. And again we will find mappings to S^2 playing a key role.

Recall from Chap. 8 that a hyper-Kähler manifold $\mathfrak{M}$ must have three independent complex structures $\{I, J, K\}$ which obey Hamilton's quaternion formula:

$$I^2 = J^2 = K^2 = -1.$$

Moreover, any linear combination, $aI + bJ + cK$, is also a Kähler structure on $\mathfrak{M}$ if $a^2 + b^2 + c^2 = 1$, which defines the unit sphere S^2 in $\mathfrak{R}^3$ with parameters $\{a, b, c\}$ [Lindstrom and Rocek, 2010].

Thus S^2 parametrizes the complex structures on $\mathfrak{M}$; and the twistor space $\mathfrak{Z}$ of the hyper-Kähler space $\mathfrak{M}$ can be defined as:

$$\mathfrak{Z} = \mathfrak{M} \otimes S^2.$$

Note that, in general, $\mathfrak{M}$ is a hyper-Kähler manifold of (real) $4n$ dimensions. Here we are considering the special case of $\mathfrak{M}$ as an ALE space (which is ADE classified), so that $\mathfrak{M}$ is a (real) four-dimensional space. Thus $\mathfrak{Z}$ is (real) six-dimensional, and so must be regarded as projective twistor space $\mathcal{PT}$, which is a complex 3D space (rather than the full twistor space $\mathcal{T} = \mathbb{C}^4$).

The relationship between $\mathcal{PT}$ and $\mathfrak{M}$ is very close. As described in Chap. 8, "ALE Spaces and Gravitational Instantons," $\mathfrak{M}$ is the desingularized version of the orbifold $\mathbb{C}^2/m$, where m is a finite subgroup of $\mathcal{SU}(2)$. Thus since m is ADE classified, then the hyper-Kähler manifolds $\mathfrak{M}$ are ADE classified. Indeed the desingularization of $\mathbb{C}^2/m$ is accomplished by blowing up the one singular point in $\mathbb{C}^2/m$ into a bouquet of copies of S^2 which intersect each other in the pattern of the corresponding ADE Coxeter graph. Thus $\mathfrak{M}$ is a complex 2D space with a smooth underlying real 4D manifold. Indeed $\mathfrak{M}$ is an asymptotically locally Euclidean (ALE) space.

The relationship between $\mathfrak{M}$ and its twistor space $\mathcal{PT}$ is clarified by the fact that $\mathbb{C}^2/m$ is the zero set of the ADE catastrophe germ (cf. Chap. 7). This germ is a harmonic polynomial as a function of three invariant polynomials $\{X, Y, Z\}$ of m, whose degrees $\{p, q, r\}$ play a fundamental role in the twistor structure $\mathcal{PT}$.

Just as $\mathbb{C}^2/m$ is the zero set of the germ $g(X, Y, Z)$, so $\mathcal{PT}$ is the zero set of $g'(X, Y, Z, \lambda)$, where $\{X, Y, Z\}$ are the m invariants, and λ is the parameter on $\mathbb{C}$ as the conformal mapping space of S^2. This is the basis of the claim that:

$$\mathcal{PT} = \mathfrak{Z} = \mathfrak{M} \otimes \mathrm{S}^2.$$

Moreover, we can see that $\mathcal{PT}$ is embedded as a complex 3D space in $\mathbb{C}^4$, with parameters $\{X, Y, Z, \lambda\}$, just as in $\mathbb{C}^2/m$ is a complex 2D space embedded in $\mathbb{C}^3$, with parameters $\{X, Y, Z\}$.

Thus there is a holomorphic fiber bundle projection:

$$\pi\colon \mathcal{PT} \to S^2$$

such that the twistor lines are sections of this fibration [Kronheimer, 1989].

In more detail, we can say that $\mathcal{PT}$ is the 3D complex hypersurface embedded in the rank-3 vector bundle over S^2:

$$\mathcal{O}(\mathrm{p}) \oplus \mathcal{O}(\mathrm{q}) \oplus \mathcal{O}(\mathrm{r}) \to S^2,$$

where $\{p, q, r\}$ are the degrees of the three invariant homogeneous polynomials of m, and each $\mathcal{O}(n)$ is a line bundle on S^2 [Dunajski and Mason, 2003].

Moreover, since n is the first Chern class of $\mathcal{O}(n)$, this means that there are n twists on the sections of the bundle $\mathcal{O}(n) \to S^2$.

To clarify the concept of a fiber-bundle twist, we note the contrast between the topologies of a cylinder and a Möbius band. A cylinder can be described as an untwisted line bundle over a circle, whereas a Möbius band is a line bundle over a circle with a single twist in the bundle structure. These contrasting topologies can be formalized as:

$$\text{Cylinder: } \mathcal{O}(0) \to S^1$$
$$\text{Möbius band: } \mathcal{O}(1) \to S^1.$$

The overall degree of the ADE catastrophe germ $g(x, y, z)$ with zero set $\mathbb{C}^2/\mathfrak{m}$ is:

$$s = p + q + r - 2.$$

Thus we also have the bundle projection:

$$\mathcal{O}(p) \oplus \mathcal{O}(q) \oplus \mathcal{O}(r) \to \mathcal{O}(s).$$

For example, in the E_7 case, $\mathbb{C}^2/\mathcal{OD}$ is the zero set of:

$$X^3 + XY^3 + Z^2,$$

where $\{X, Y, Z\}$ are polynomials of degrees $\{12, 8, 18\}$. The overall degree of this (weighted homogeneous) polynomial is 36, since:

$$X^3 \text{ has total degree } 3(12) = 36$$
$$XY^3 \text{ has total degree } 12 + 3(8) = 36$$
$$Z^2 \text{ has total degree } 2(18) = 36.$$

Note that this corresponds to $12 + 8 + 18 - 2 = 36$. Also note that 36 is 2(18), where 18 is the Coxeter number of the E_7 Coxeter group.

In general, for each ADE classified twistor space, the highest line-bundle twist is the highest degree of the McKay group invariant polynomial, and is also the corresponding ADE Coxeter number of the Coxeter graph. More explicitly, we can display the following table of ADE structures (where we have used the conventions of Chap. 7):

Graph label	McKay group	Germ of ADE catastrophe	Degrees of invariants (y, x, z)	Degree of the germ 2(Coxeter #)
A_k	$\mathcal{Z}_{k+1}$	$x^{k+1} + y^2 + z^2$	$2, k+1, k+1$	$2(k+1)$
D_k	$\mathcal{Q}_{k-2}$	$x^2y + y^{k-1} + z^2$	$4, 2(k-2), 2(k-1)$	$4k - 4$
E_6	$\mathcal{TD}$	$x^3 + y^4 + z^2$	6, 8, 12	24
E_7	$\mathcal{OD}$	$x^3 + xy^3 + z^2$	8, 12, 18	36
E_8	$\mathcal{ID}$	$x^3 + y^5 + z^2$	12, 20, 30	60

Note that the degree of the germ is twice the degree of the highest invariant (which is the ADE Coxeter number). That is:

$$p + q + r - 2 = s = 2(r).$$

Moreover, the twistor space $\mathcal{PT}$ as the zero set of $g'(x, y, z, \lambda)$, has an embedding into the rank-3 vector bundle:

$$\mathcal{O}(p) \oplus \mathcal{O}(q) \oplus \mathcal{O}(r) \to S^2.$$

This implies that there are three separate projections of $\mathcal{PT}$ onto the line bundles, where $\{y, x, z\}$ are as in the table above:

$$\mathcal{PT} \to \mathcal{O}(p) \text{ since } y(\lambda) \in \mathcal{O}(p)$$
$$\mathcal{PT} \to \mathcal{O}(q) \text{ since } x(\lambda) \in \mathcal{O}(q)$$
$$\mathcal{PT} \to \mathcal{O}(r) \text{ since } z(\lambda) \in \mathcal{O}(r).$$

Note that, here, y is the McKay group invariant of lowest degree and thus the easiest to describe. The projection onto the y coordinate in S^2:

$$y(\lambda) \in \mathcal{O}(p)$$
$$\text{in } \mathcal{PT} \to \mathcal{O}(p)$$

has fibers which are affine conics in the $\mathcal{A}_r$ and $\mathcal{D}_r$ cases, and affine elliptic curves in the E_6, E_7 and E_8 cases.

Thus every ADE classified twistor space $\mathcal{PT}$ can be realized as an affine line bundle over $\mathcal{O}(p)$. A much more detailed description of this construction can be found in [Dunajski and Mason, 2003].

Since $\mathcal{PT} = \hbar\mathcal{K} \otimes S^2$ (where $\hbar\mathcal{K}$, a hyper-Kähler space, is an ADE classified gravitational instanton), it may not be too surprising to find S^2 playing a role in the relationship between twistors and supersymmetry (and thus supergravity). Therefore the Penrose conjecture that twistors lead to quantum gravity would find some fulfillment.

Supersymmetry (SUSY) was introduced in 1970 as a transformation that changes bosons into fermions (and vice versa). This was necessary to provide a way for string theory to contain fermionic states in addition to the bosonic states of the original 26D spacetime string theory.

Superstring theory (of necessity obeying SUSY) was a 10D spacetime theory, which began as a model for the strong nuclear force. However, a major embarrassment (other than the hyperdimensionality of spacetime) was the necessary inclusion of a spin-2 massless boson in the spin spectrum. No such particle could be found in the high energy realm of the nuclear forces.

The graviton, of course, had long been predicted as the quantum of gravity; and it had to be a massless spin-2 boson. This suggested that the focus of SUSY should be on the low energy realm of gravity, rather than on the high energy nuclear realm [Kaku, 1999; Green, Schwarz and Witten, 1987].

Thus from SUSY in the early '70s two approaches to quantum gravity were developed: supergravity and superstring theory.

Superstring theory was described in some detail in Chap. 8: "ALE spaces and gravitational Instantons."

Here we will focus on the point-particle limit of superstring theory, which is supergravity [van Nieuwenhuizen and Freedman, 1979]. The great excitement about SUSY was that it could be considered as a gauge theory of gravity, analogous to the gauge theories of the $\mathcal{U}(1) \times \mathcal{SU}(2) \times \mathcal{SU}(3)$ Standard Model gauge groups. Indeed, if SUSY is applied locally to spacetime (rather than the

SUSY transformations being applied to all points of spacetime globally), then the field equations of Einstein's general relativity would be obeyed. Indeed, general relativity can be regarded as ($N = 0$) supergravity, while supergravities with $N > 0$ would be regarded as extensions of general relativity [van Nieuwenhuizen, 1980].

Moreover, it can be shown that $N = 2$ supergravity unifies gravity with electromagnetism (a fond dream of Einstein's) [Kuzenko, 2012].

Note that although supergravity can be formulated in spacetime dimensions up to 11D, we will consider here only the 4D spacetime version. In this case supergravity is formulated on a superspace with four bosonic dimensions, and $4N$ fermionic dimensions. These fermionic dimensions are parametrized by Grassmann numbers, which are anticommutative, whereas the bosonic dimensions are parametrized by real numbers (which, of course, are commutative). The $4N$ Grassman numbers also correspond to the $4N$ supercharges, where N is the number of supersymmetry transformations. Thus $N = 2$ supergravity would have eight supercharges. See Appendix A4 in [Kaku, 1999].

We note in passing that the main reason for going beyond supergravity to superstring theory is the promise of unifying all the forces. Also, as is emphasized in this book, we are enthralled by the vast mathematical vistas revealed by this adventure.

Here we will describe the mappings:

$$\begin{array}{ccc} \text{Supersymmetry} & \rightarrow & \text{Superspace} \\ \downarrow & & \downarrow \\ \text{Twistors} & \rightarrow & \text{hyper-Kähler ALE spaces.} \end{array}$$

As already mentioned, twistors ($\mathcal{PT}$) are mapped to hyper-Kähler space $\hbar\mathcal{K}$ via:

$$\mathcal{PT} = \hbar\mathcal{K} \otimes S^2,$$

where the three complex structures (I, J, K) on $\hbar\mathcal{K}$ can be smoothly rotated by the parameters of S^2. The complex structures are also closely related to the three symplectic structures on $\hbar\mathcal{K}$, which is what makes $\hbar\mathcal{K}$ a hyper-Kähler manifold. One possible mapping is

[Lindstrom and Rocek, 2010]:

$$J \to \omega^{(1,1)}$$
$$I \to \omega^{(2,0)}$$
$$K \to \omega^{(0,2)}.$$

Thus we can combine these symplectic 2-forms into:

$$\Omega(\lambda) = \omega^{(2,0)} + \lambda\omega^{(1,1)} - \lambda\omega^{(0,2)},$$

where λ is a parameter on S^2, so that $\Omega(\lambda)$ is a section of the 2-form valued bundle:

$$\mathcal{O}(2) \to S^2.$$

It is perhaps surprising that this same S^2 considered as a Riemann sphere in the twistor space $\mathcal{PT}$ plays the same role in the $N = 2$ superspace (with four bosonic dimensions and eight fermionic dimensions). The fields on this space are called superfields, and they transform according to the analytic properties of S^2 as parametrized by λ.

Note that although this geometry is called the projective superspace construction, the projectivity here is due to the conformal mapping of $\mathbb{C}$ to S^2 (also called the "stereographic" projection).

For example, the scalar superfield is called the "arctic" multiplet Υ, because it is analytic around the North pole of S^2, while the "antarctic" multiplet Υ' is analytic around the South pole. Since these two polar regions are covered by separate coordinate patches, the overlap region corresponds to "tropical" superfields.

Due to the ADE classification of the ALE spaces $\hbar\mathcal{K}$ in the overall space, $\mathcal{PT} = \hbar\mathcal{K} \otimes S^2$, we can define functions of the superfields which play a fundamental role in the dynamics of this geometry. These special functions are called superpotentials. They are indeed ADE classified since they are exactly equivalent to the germs of the ADE catastrophes, as listed above. For example, the E_7 superpotential is:

$$\mathcal{W}(\mathrm{E}_7) = x^3 + xy^3 + z^2,$$

where $\{x, y, z\}$ are scalar functions of $\mathbb{C}^2$.

We are still in the context of $N = 2$ supersymmetry, so that in a formal sense we can write the general definitions of the two generators of $N = 2$ supersymmetry as:

$$Q_1 = \frac{1}{2}[(p - i\mathcal{W})b + (p + i\mathcal{W})b^*],$$
$$Q_2 = \frac{i}{2}[(p - i\mathcal{W})b - (p + i\mathcal{W})b^*],$$

where p is the momentum of a particle so that the symplectic formula, $[x, p] = i$, is in accord with the symplectic structure on $\hbar\mathcal{K}$ as described above.

Also b and b^* obey the Grassmann relations:

$$\{b, b^*\} = 1; \quad b^2 = 0.$$

Note that Q_1 and Q_2 are supersymmetry operators in the sense that they transform bosonic states into fermionic states and vice versa.

Thus the Hamiltonian operator can be abstractly given as:

$$H = \frac{1}{2}[p^2 + \mathcal{W}^2 \pm \partial\mathcal{W}].$$

Moreover, by way of the superpotentials $\mathcal{W}$, we make contact with the ADE classification of 2D conformal field theory, which will be the subject of the next chapter.

Chapter 11

Two-Dimensional Conformal Field Theories

In string theory, the strings sweep out a 2D worldsheet, which is the string analog to the world-lines swept out by the particles (as pictured so familiarly in the Feynman diagrams). This 2D worldsheet is embedded in the 10D spacetime of any of the five superstring theories (cf. Chap. 8).

We note in passing that the string interactions are a loop (or open string) version of the point particle interactions of Feynman's quantum field theory. Thus the infinities of point-particle interactions disappear along with the necessity to cancel these infinities by the mathematical procedure called renormalization [Witten, 1996].

The fields which live on the string worldsheet correspond to the embedding of the worldsheet in spacetime. These fields obey several closely related symmetries: reparametrization invariance; 2D Lorentz invariance; supersymmetry; and 2D conformal invariance (a kind of master invariance) which is the subject of this chapter.

In Chap. 10 we have met with the twistor conformal group, $\mathcal{SU}(2,2)$ acting on the twistor space $\mathbb{C}^4$. Here we will encounter the group $\mathcal{SO}(2,2) = \mathcal{SL}(2\ \mathfrak{R})_{\text{left}} \otimes \mathcal{SL}(2\ \mathfrak{R})_{\text{right}}$. This is a finite-dimensional subgroup of the infinite dimensional conformal group on the 2D worldsheet.

The conformal group (on any manifold) is the set of all conformal mappings of the manifold. Any conformal mapping preserves angles (and therefore direction), but it does not preserve distance. Thus changes in scale do not destroy the validity of a conformally invariant theory. The best simple example of a conformal mapping is the "stereographic" mapping of a 2D plane to a 2D sphere, or the reverse mapping. If the plane is considered the Argand plane of complex numbers $\mathbb{C}$, then the 2D sphere is called the Riemann sphere. A point at infinity of $\mathbb{C}$ is mapped to the "north" pole of the Riemann sphere, the "south" pole being mapped to the origin point of $\mathbb{C}$.

The Riemann sphere (plus a point at infinity) conformally maps all of $\mathbb{C}$ and is the simplest example of a Riemann surface. However, subregions of $\mathbb{C}$ are also Riemann surfaces.

The string worldsheet consists of a set of segments, each of which is a Riemann surface of genus g, where g is the number of holes. These holes are the worldsheet analog of the loops in the Feynman diagrams of particle physics. Each of these worldsheet segments can be conformally mapped to a punctured Riemann sphere, where the punctures correspond to the worldsheet's topological holes.

The set of all conformal mappings on the worldsheet forms a 2D conformal group, which (as an infinite dimensional group) is the symmetry group of 2D conformal field theory (CFT).

This 2D CFT is related to the worldsheet supersymmetry. Moreover, since the 2D worldsheet is embedded in 10D spacetime, there is also a 10D spacetime supersymmetry. The discovery of the relationship between the conformal invariance of the 2D worldsheet and the supersymmetry of the 10D spacetime was a major turning point in the development of string theory.

There are two different formalisms for the supersymmetry:

(1) The Raymond–Neveu–Schwarz (RNS) formalism models the supersymmetry of the 2D worldsheet.
(2) The Green–Schwarz (GS) formalism models the supersymmetry of the 10D spacetime.

These two formalisms are bridged by conformal field theory. In fact, the superconformal symmetry of the RNS string entails constraints that fix the ten-dimensionality of the embedding spacetime [Kaku, 1999; Becker, Becker and Schwarz, 2007].

CFT is also entailed in the holographic principle as exemplified by the AdS/CFT duality in which $\mathcal{N} = 4$ super Yang–Mills theory lives on the 4D boundary of the 5D anti-de Sitter space in the 10D spacetime ($\mathrm{AdS}_5 \otimes S^5$) of the IIB superstring theory. Cf. Chap. 8.

Here, however, we will concentrate on 2D superconformal field theory. In fact we will specialize to minimal $\mathcal{N} = 2$ superconformal field theory. A substructure within this theory, called rational conformal field theory (RCFT), will yield a rather surprising ADE classification of these sub-theories [Zuber, 2002].

Note that we have already dealt with the five superstring theories and their relationship to the ADE classification of the ALE space in Chap. 8. Although not mentioned in that chapter, it is significant that the supersymmetry of the 2D worldsheet is a consequence of superconformal invariance.

Also note that the 26D spacetime of the bosonic string theory is derived from the conformal invariance of the 2D bosonic worldsheet [Polchinski, 1998, Vol. I].

Moreover, the two heterotic string theories are a profound interweaving of the bosonic 26D theory and the supersymmetric 10D theories. This is accomplished by the fact that in the 2D worldsheet left-moving strings see the entire 26D spacetime, whereas the right-movers see only the 10D spacetime. This is possible only because in a 2D conformal theory, a single boson can be constructed from a combination of a fermion and anti-fermion. This formalism is called bosonization. In order for the quantum anomalies to be cancelled, the 16D mismatch between the 26D bosonic theory and the 10D superstring theory must be parametrized by a lattice corresponding to Coxeter graphs of rank 16. The two possible cases are:

D_{16} the label for the Lie group $\mathcal{SO}(32)/\mathcal{Z}_2$

$\mathrm{E}_8 \oplus \mathrm{E}_8$

where, in each case, the Lie algebra is 496-dimensional, with a 16-dimensional Cartan subalgebra.

Moreover, the standard model algebras can be embedded in $\mathrm{E}_8 \oplus \mathrm{E}_8$ by the ADE series of projections (cf. Chap. 5):

$$\mathrm{E}_8 \to \mathrm{E}_6 \to \mathrm{D}_5 \to \mathrm{A}_4.$$

Thus we can recognize the grand unified theory (GUT) Lie groups:

$$\mathrm{D}_5\ (\mathcal{SO}(10)) \quad \text{and}$$

A_4 ($\mathcal{SU}(5)$), whose maximal subgroup is $\mathcal{U}(1) \otimes \mathcal{SU}(2) \otimes \mathcal{SU}(3)$, the three gauge groups for the electrical force, the weak force, and the strong color force (as in quantum chromodynamics). Note also that the three Standard Model gauge groups correspond to A_0, A_1 and A_2 [Kaku, 1993].

We take up now the quite different ADE classification of the RCFT within the minimal $\mathcal{N} = 2$ superconformal theory. The term "minimal" here means that we are describing the $\mathcal{N} = 2$ superstring theory with the 2D string worldsheet embedded in 10D spacetime.

A "non-minimal" theory would omit the requirement of this 10D embedding, and this would correspond to a much expanded classification scheme not restricted to the ADE series [Gray, 2012].

For the ADE classification considered here, the master algebra is the **sl**(2) affine Kac–Moody algebra, which is infinite dimensional. This corresponds to the fact that the conformal group on 2D space is infinite dimensional.

Note that the ordinary **sl**(2) Lie algebra (as a three-dimensional subalgebra of the affine Kac–Moody version) has the label A_1, so that its Coxeter graph is a single node (cf. Chap. 5). The compact form of the A_1 Lie algebra is **su**(2), which generates the Lie group $\mathcal{SU}(2)$.

Since we have described the ADE classification of the finite subgroups of $\mathcal{SU}(2)$ in Chap. 4 (the McKay groups), it is not too surprising to find that the integrable representations of the affine **sl**(2) are ADE classified. Remember that the McKay correspondence entails the relationship of the ADE affine Kac–Moody algebras to the ADE classified McKay groups, i.e. finite subgroups of $\mathcal{SU}(2)$.

For the bosonic string, the conformal invariance of the 2D worldsheet is formalized by the infinite dimensional Virasoro algebra, which in its quantum version has the central extension, whose constant c is called the central charge:

$$[L_m, L_n] = (m - n)L_{m+n} + (c/12)(m^3 - m)\delta_{m+n,0}.$$

Here the central charge constant c is also called the conformal anomaly, because it creates an inconsistency for the quantum version of this theory. However, by fixing the spacetime dimension at 26, the anomaly is canceled (i.e. c = 0), and the theory implies Poincaré invariance in the 26D spacetime. Thus the bosonic string theory is described as a theory of strings sweeping out a 2D spacetime in an underlying 26D spacetime [Polchinski, Vol 1, 1998].

However, the superstring theory entails a very significant extension of the bosonic Virasaro algebra, called the super Virasoro algebra. In this case, there is at least one supercharge generator. The most useful cases are the addition of 1 or 2 supercharge generators, corresponding to $\mathcal{N} = 1$ and $\mathcal{N} = 2$ superstring theories. The highest possible case is $\mathcal{N} = 8$.

The $\mathcal{N} = 2$ superconformal algebra is an infinite dimensional algebra formalized as the bosonic string ($\mathcal{N} = 0$) relations as well as a much expanded set of relations. Note that in the following formalism $[X, Y] = XY - YX$; and $\{X, Y\} = XY + YX$. Also note that $r, s \in Z$ yields the Raymond algebra, whereas $r, s \in 1/2 + Z$ yields the Neveu–Schwarz algebra [Polchinski, Vol II, 1998].

$$
\begin{aligned}
&[L_m, L_n] = (m-n)L_{m+n} + (c/12)(m^3 - m)k\delta_{m+n,0},\\
&[L_m, J_n] = -nJ_{m+n}; \quad [J_m, J_n] = (c/3)m\delta_{m+n,0},\\
&\{G_r^+, G_s^-\} = 2L_{r+s} + (r-s)J_{r+s} + (c/3)(r^2 - 1/4)\delta_{r,-s},\\
&\{G_r^+, G_s^+\} = \{G_r^-, G_s^-\} = 0.\\
&[L_m, G_r^+] = ((m/2) - r)G_{r+m}^+; \quad [L_m, G_r^-] = ((m/2) - r)G_{r+m}^-.\\
&[J_m, G_r^+] = G_{m+r}^+; \quad [J_m, G_r^-] = -G_{m+r}^-.
\end{aligned}
$$

In this $\mathcal{N} = 2$ superconformal algebra, the J operators are called "current" operators, because they correspond to an early form of supersymmetry operators forming a current algebra. Also the two supersymmetry operators here are:

$$
\begin{aligned}
G_r^- &= \sum (a_{r+m} - ib_{r+m}) - e_m^*\\
G_r^+ &= \sum (a_{-m} + ib_{-n}) \cdot e_{r+m},
\end{aligned}
$$

where e_m^* is the complex conjugate of e_m. Also the central charge $c = 3$ in this formalism, which corresponds to the $\mathcal{N} = 2$ Neveu–Schwarz algebra.

There is, however, an infinite number of such $\mathcal{N} = 2$ superconformal algebras corresponding to different levels k of the affine $\mathbf{sl}(2)$ Kac–Moody algebra. As we will see, these levels correspond to an ADE classification of these $\mathcal{N} = 2$ superconformal algebras.

This affine **sl**(2) algebra at level $\mathscr{k}$ and with basis structure, E_n, F_n, H_n, obeys the algebraic formalism:

$$[H_m, H_n] = 2m\mathscr{k}\delta_{m+n,0},$$
$$[E_m, F_n] = H_{m+n} + m\mathscr{k}\delta_{m+n,0},$$
$$[H_m, E_n] = 2E_{m+n},$$
$$[H_m, F_n] = -2F_{m+n}.$$

The two supersymmetry generators (required by the $\mathcal{N} = 2$ theory) are:

$$G_r^+ = (\mathscr{k}/2 + 1)^{-1/2}\Sigma E_{-m}(e_{m+r}),$$
$$G_r^- = (\mathscr{k}/2 + 1)^{-1/2}\Sigma F_{-m}(e_{m^*}).$$

So that the super Virasaro algebra has central charge:

$$c = 3\mathscr{k}/(\mathscr{k} + 2).$$

It is useful to clarify certain group theory structures. The strings sweeping out the 2D worldsheet are of two basic types — where for convenience we assume a Euclidean metric with signature (+, +), which we interpret as applying to $\mathbb{C}$. Thus (for $z \in \mathbb{C}$, and for z^* conjugate to z, and L_n conjugate to $\underline{L}_n$) the generators of the conformal group are:

$$L_n = -z^{n+1}\partial/\partial_z \quad \text{and} \quad \underline{L}_n = -z^{n+1}\partial/\partial_{z^*}.$$

Open strings are simply lines of extremely small finite length. The 2D worldsheet is invariant under the $\mathcal{SL}(2, \mathfrak{R})$ group generated by the **sl**(2, $\mathfrak{R}$) algebra. This is a Virasoro subalgebra, because it is a three-dimensional algebra with the three basis elements:

$$L_0, L_1, L_{-1}.$$

Note that the compact form of **sl**(2, $\mathfrak{R}$) is **su**(2,$\mathcal{C}$). Moreover, just as the Lie groups $\mathcal{SU}(2)$ and $\mathcal{SO}(3)$ have the same Lie algebra so that the Lie groups $\mathcal{SL}(2, \mathfrak{R})$ and $\mathcal{SO}(2, 1)$ are generated by the same Lie algebra **sl**(2, $\mathfrak{R}$). These Lie groups and Lie algebras all correspond to the A_1 Coxeter graph (also called the A_1 Dynkin diagram). Cf. Chap. 5.

Closed strings are extremely small circles, which sweep out 2D tubular versions of the point-particle world lines. Because these strings are closed and oriented, they can rotate either clockwise or counterclockwise, and are called right movers and left movers, respectively. In this case the subalgebra of the Virasoro algebra is:

$$\mathbf{sl}(2,\mathfrak{R})_{\text{left}} \oplus \mathbf{sl}(2,\mathfrak{R})_{\text{right}}$$

(with basis elements): L_0, L_1, L_{-1} and $\underline{L}_0, \underline{L}_1, \underline{L}_{-1}$.

Here we note that this Lie algebra sum is equivalent to the Lie algebra $\mathbf{so}(2, 2)$, which generates the Lie group $\mathcal{SO}(2,2)$. This group, as mentioned above, is an analog to the twistor conformal group $\mathcal{SU}(2,2)$ acting on the twistor-space $\mathbb{C}^4$.

Clearly the $\mathbf{sl}(2, \mathfrak{R})$ algebra is a key to the 2D conformal field structure. It is the largest finite-dimensional Lie subalgebra of the infinite-dimensional affine $\mathbf{sl}(2,\mathfrak{R})$ Kac–Moody algebra. Moreover, the $\mathcal{SL}(2,\mathfrak{R})$ Lie group, generated by $\mathbf{sl}(2,\mathfrak{R})$, has the very important finite subgroup $\mathcal{SL}(2,\mathcal{Z})$, where $\mathcal{Z}$ is the set of integers. This group is also called the modular invariance group, and as such relates directly to the ADE classification of the 2D CFTs [Zuber, 2002].

The ADE classification of the 2D CFTs entails the classification of 2D RCFTs. The term "rational" here means that the various mathematical constants entailed by the RCFT are rational numbers. These theories obey the modular invariance group $\mathcal{SL}(2,\mathcal{Z})$. The 2D representation space is mapped onto a torus by identifying the opposite edges of an elementary 4-sided trapezoidal cell in the 2D complex plane.

For the complex number τ, the modular group $\mathcal{SL}(2,\mathcal{Z})$, can be defined by the transformation:

$$\tau \to \frac{a\tau + b}{c\tau + d}$$

where $\{a, b, c, d\}$ are integers such that the determinant is $ad - bc = 1$. Thus the modular group $\mathcal{SL}(2,\mathcal{Z})$ is generated by the transformations:

$$T\colon \tau \to \tau + 1 \quad \text{and} \quad S\colon \tau \to -1/\tau.$$

Therefore, $\mathcal{SL}(2,\mathcal{Z})$ is defined by the relations:

$$S^2 = 1 \quad \text{and} \quad (ST)^3 = 1.$$

Since the modular group is defined with reference to the 2D torus, we can describe a partition function as a path integral on this torus.

These path integrals are analogs to the Feynman path integrals of quantum field theory.

In this RCFT context, there will be an infinite number of these partition functions, each corresponding to an irreducible representation of the modular group as a structure embedded in the infinite dimensional affine $\mathbf{sl}(2,\mathfrak{R})$ Kac–Moody algebra. These partition functions correspond to the levels k of this affine $\mathbf{sl}(2,\mathfrak{R})$, and are composed of polynomials over the characters χ_λ of the of this affine $\mathbf{sl}(2,\mathfrak{R})$.

In this case, complete calculations have been made and can be displayed explicitly. In the following table, the ADE classification of these partition functions is clearly displayed [Cappelli and Zuber, 2009; Gannon, 2000].

The ADE classification of the affine sl(2) partition functions.

Level	Partition function	Label
$k \geq 0$	$\sum_{\lambda=1\to k+1} \lvert\chi_\lambda^2\rvert$	A_{k+1}
$k = 4\rho \geq 4$	$\sum_{\lambda=1\,\text{odd}-1\to 2\rho-1} \lvert\chi_\lambda + \chi_{4\rho+2-\lambda}\rvert^2 + 2\lvert\chi_{2\rho+1}\rvert^2$	$D_{2\rho+2}$
$k = 4\rho - 2 \geq 6$	$\sum_{\lambda\,\text{odd}-1\to 2\rho-1} \lvert\chi_\lambda\rvert^2 + \lvert\chi_{2\rho}\rvert^2 + \sum_{\lambda\,\text{even}-2\to 2\rho-2} (\chi_\lambda\chi_{4\rho-\lambda}^* + \text{c.c.})$	$D_{2\rho+1}$
$k = 10$	$\lvert\chi_1 + \chi_7\rvert^2 + \lvert\chi_4 + \chi_8\rvert^2 + \lvert\chi_5 + \chi_{11}\rvert^2$	E_6
$k = 16$	$\lvert\chi_1 + \chi_{17}\rvert^2 + \lvert\chi_5 + \chi_{13}\rvert^2 + \lvert\chi_7 + \chi_{11}\rvert^2 + [\lvert\chi_9\rvert^2 + (\chi_3 + \chi_{15})\chi_9^* + \text{c.c.}]$	E_7
$k = 28$	$\lvert\chi_1 + \chi_{11} + \chi_{19} + \chi_{29}\rvert^2 + \lvert\chi_7 + \chi_{13} + \chi_{17} + \chi_{23}\rvert^2$	E_8

Note that the central charge in each case is $c = 3k/k+2$, which is ≤ 3. Thus for example: $c(E_6) = 2.5$; $c(E_7) = 2.6666\ldots$; $c(E_8) = 2.8$.

Note also that in each case $k + 2$ is the Coxeter number K of the corresponding ADE Coxeter graph. Cf. Chap. 6.

Perhaps the most remarkable aspect of these ADE partition functions, is that the index λ of the characters χ_λ corresponds almost exactly to the Coxeter exponents, where the number of exponents is the rank of the Coxeter graph. The exception is to be found in the

E_7 case, where we find 3 and 15 as well as the E_7 Coxeter exponents: 1, 17, 5, 13, 7, 11, 9.

Note that each Coxeter exponent plus 1 is equal to a fundamental degree of the Coxeter group. The order of this group is equal to the product of these fundamental degrees (cf. Chap. 6).

For example, the order of the E_7 Coxeter group is 2,903,040, which is the product of the E_7 fundamental degrees: 2,6,8,10,12,14,18.

We can display the following table:

ADE table of Coxeter exponents.

Label	Coxeter #: $K = k+2$	Exponents
A_n	$n+1$	$1, 2, \ldots, n+1$
D_{n+2}	$2(n+1)$	$1, 3, \ldots, 2n+1, n+1$
E_6	12	$1, 4, 5, 7, 8, 11$
E_7	18	$1, 5, 7, 9, 11, 13, 17$
E_8	30	$1, 7, 11, 13, 17, 19, 23, 29$

Note: In each case, the Coxeter number K is 1 + the highest exponent.

Moreover, we can derive the dimensionality $\mathcal{D}$ of the ADE Lie algebras by the simple formula:

$$\mathcal{D} = nK + n,$$

where the rank n is the dimensionality of the Cartan subalgebra.

Also we can derive $\mathcal{D}$ from the ADE exponents $\mathcal{E}$:

$$\mathcal{D} = \sum_{1}^{n} 2\mathcal{E} + 1.$$

For example, the E_7 Lie algebra has:

$$\mathcal{D} = 3 + 11 + 15 + 19 + 23 + 27 + 35 = 133.$$

Geometrically, these numbers correspond to the fact that the compact form of the E_7 Lie group is the direct product of seven spheres:

$$S^3 \otimes S^{11} \otimes S^{15} \otimes S^{19} \otimes S^{23} \otimes S^{27} \otimes S^{35}$$

while the Cartan subgroup $\mathcal{T}^7$ consists of the product of seven circles, one projected from each of these seven spheres.

Analogous calculations can be made for each ADE Lie algebra [Hiller, 1982].

There is also another ADE classification for the minimal $\mathcal{N} = 2$ superconformal field theories (in 2D spacetime). This entails Landau–Ginzburg models of superpotentials $\mathcal{W}$ whose critical points are stationary (even at the quantum level) because of supersymmetry. This ADE classification corresponds exactly to the germs of the Arnold–Thom catastrophe structures as described in Chap. 7. These germs in the present context are polynomial functions of chiral fields.

Here we will display the full set of ADE catastrophe germs, including the quadratic term. This quadratic term is necessary to make contact with the ADE classification of ALE spaces as described in Chap. 8. These ADE spaces are the desingularized form of the orbifold $\mathbb{C}^2/m$, where m is a finite subgroup of $\mathcal{SU}(2)$, and are thus ADE classified (cf. Chap. 4). In each ADE case, $\mathbb{C}^2/m$ is the zero set of the corresponding polynomial function of chiral fields, $\mathcal{W}(X, Y, Z)$.

Since the $\mathcal{N} = 4$ superconformal field theory entails the ADE classification of ALE space, we see that $\mathcal{N} = 2$ superconformal field theory is nested within the $\mathcal{N} = 4$ theory by way of the ADE classification of the superpotentials $\mathcal{W}(X, Y)$ within $\mathcal{W}(X, Y, Z)$.

ADE classification of superpotentials $\mathcal{W}(X, Y, Z)$.

Label	m	$\lvert m \rvert$	$\mathcal{W}$
A_{n-1}	$\mathcal{Z}_n$	n	$X^n - YZ$
D_{n+2}	$\mathcal{Q}_n$	$4n$	$X^{n+1} + XY^2 + Z^2$
E_6	$\mathcal{TD}$	24	$X^4 + Y^3 + Z^2$
E_7	$\mathcal{OD}$	48	$X^3 + XY^3 + Z^2$
E_8	$\mathcal{ID}$	120	$X^5 + Y^3 + Z^2$

In this $\mathcal{N} = 4$ superconformal field theory, the modular invariant partition functions differ markedly from the $\mathcal{N} = 2$ version of these invariants, as listed above. For example in the E_7 case, we have the level $k = 16$ partition functions:

$$\begin{aligned} \mathcal{N} = 2\colon \quad & |\chi_1 + \chi_{17}|^2 + |\chi_5 + \chi_{13}|^2 + |\chi_7 + \chi_{11}|^2 \\ & +[|\chi_9|^2 + (\chi_3 + \chi_{15})\chi_9 * + \text{c.c.}] \\ \mathcal{N} = 4 \quad & |\chi_0 + \chi_8|^2 + |\chi_2 + \chi_6|^2 + |\chi_3 + \chi_5|^2 + |\chi_4|^2 \\ & +[(\chi_1 + \chi_7)^* \chi_4 + \chi_4^*(\chi_1 + \chi_7)] \end{aligned}$$

cf. [Anselmi *et al.*, 1994], which needs to be read in the context of the ADE classification of ALE spaces as in Chap. 8.

The techniques of superconformal field theories have migrated over to the more down-to-earth models of critical phenomena in solid state theory. This is possible because at a critical point such a system is scale invariant. And it happens that the conformal invariance of various critical phenomena can be ADE classified [Henkel, 1999; Cappelli and Zuber, 2009].

The most important of these critical phenomena correspond to minimal conformal models. These are the only RCFTs with central charge $c < 1$.

These minimal models are two-dimensional theories, and the ADE classification entails partition functions that correspond to pairs of Coxeter graphs, one of which is always of A-type. This ADE classification can be indexed by pairs of Coxeter labels. We can list some well-known models:

A_2, A_3	Ising model: Central charge = 0
A_3, A_4	Tricitical Ising model: Central charge = 0.5
A_{m-1}, A_m	Restricted solid-on-solid generalizations: Central charge = $c = 1 - 6/(m(m+1))$
A_4, D_4	3-state Potts model: Central charge = 0.7
A_6, D_4	Tricritical Potts model: Central charge = 0.857...

In general, all the Critical Point Partition Functions (for $c < 1$) have been calculated for the ADE series. There is yet no known physical model for higher rank D and E Coxeter graphs. As an example of one such partition function, we can display the case:

E_7, A_{18} (where m is 18) and we have a modified form of the E_7 partition function:

$$\frac{1}{2} \sum_{s=1 \to 18} [|\chi_{1,s} + \chi_{17,s}|^2 + |\chi_{5,s} + \chi_{13,s}|^2 + |\chi_{7,s} + \chi_{11,s}|^2$$
$$+ |\chi_{9,s}|^2 + (\chi_{3,s} + \chi_{15,s}) * \chi_{9,s} + \text{c.c}]$$

cf. [Henkel, 1999] for the ADE critical point partition function table.

To briefly summarize, we have described two separate ADE classifications:

(1) Minimal $\mathcal{N} = 2$ superconformal field theories (on the 2D worldsheet embedded in 10D spacetime), central charge $c < 3$.
(2) Critical phenomena in 2D space, using some methods borrowed from (1). Here the central charge is $c < 1$.

Chapter 12

Elliptic Curves and the Monster Group

In algebraic geometry, curves and surfaces (including hypersurfaces) are defined over complex numbers $\mathbb{C}$. Thus an elliptic curve would be a 2D real space (that is, a 1D complex space). This 2D real space is topologically equivalent to a 2D torus T^2. Such a torus can be constructed from a 2D lattice in $\mathbb{C}$ by identifying the opposite edges of all the parallelograms in the lattice. Any such lattice can be generated by two numbers: 1 (on the real number line) and τ in $\mathcal{H}$ the upper-half plane of $\mathbb{C}$.

Moreover, any elliptic curve can be defined as the zero set of a third degree polynomial. In the formalism of Weierstrass, this polynomial is very simple:

$$4X^3 - G_2X - G_3 - Y^2 = 0,$$

where $G_2 = 60\Sigma(m+n\tau)^{-4}$ and $G_3 = 140\Sigma(m+n\tau)^{-6}$, while assuming that neither m nor n is 0. Thus most of the complexity of this formalism is contained in G_2 and G_3, which are infinite sums.

In this context there is a rather magical function called Klein's absolute invariant, or simply the j-invariant.

This j-invariant is a modular function, since it is invariant under the modular group $\mathcal{SL}(2, \mathcal{Z})$, as described in Chap. 11. It is the unique modular function in the sense that any modular function is some rational function of j. There are special values of j:

$$j(\exp(2/3\pi i)) = 0; \quad j(i) = 1728.$$

By way of the special functions, G_2 and G_3, Felix Klein in his *Lectures on the Icosahedron* [Klein, 1956] defined his absolute invariant as:

$$J(\tau) = (G_2)^3/\Delta,$$

where Δ is the modular discriminant:

$$\Delta = (G_2)^3 - 27(G_3)^2$$

so that the j-invariant is defined as:

$$j(\tau) = 1728(G_2)^3/\Delta = 1728\ J(\tau)$$

and we note that 1728 gives the j-invariant an integral Fourier expansion.

However, the most amazing property of the j-invariant is that it corresponds to the Monster group, in the sense that the integers of the Fourier expansion of j relate to the irreducible representations of the Monster group [Conway and Sloane, 1988; Borcherds, 2002].

This Fourier expansion of j is with respect to $q = \exp(2\pi i\tau)$, where τ is in the complex upper-half plane $\mathcal{H}$. This expansion begins with:

$$\begin{aligned} j(\tau) &= 1/q + 744 + 196884q + 21493760q^2 + 864299970q^3 \\ &\quad + 20245856256q^4 + 333202640600q^5 + 4252023300096q^6 + \cdots . \end{aligned}$$

John McKay in 1978 noticed that 196,884 was equal to 1+196,883, whereas the Monster group (not yet proved to exist) had its smallest (non-trivial) representation as transformations of a 196883-dimensional vector space [Ronan, 2006].

John Thompson took this coincidence seriously enough to point out further Monster group irreducible representation sums:

$$\begin{aligned} 196884 &= 196883 + 1, \\ 21493760 &= 21296876 + 196883 + 1, \\ 864299970 &= 842609326 + 21296876 + (2)196883 + (2)1, \\ 20245856256 &= 18538750076 + (2)842609326 + 21296876 \\ &\quad +(3)196883 + (3)1. \end{aligned}$$

Moreover, Thompson could break up 20,245,856,256 as a different sum of Monster group irreducible representations:

$$\begin{aligned} 20245856256 &= 1936002527 + 842609326 + (2)21296876 \\ &\quad +(3)196883 + (2)1. \end{aligned}$$

Thus McKay and Thompson suggested that there must exist an infinite-dimensional graded representation of the Monster group $\mathfrak{M}$, where each grade would correspond to a vector space direct sum over representation orders of $\mathfrak{M}$.

For this purpose the j-invariant is transformed to:

$$j(\tau) - 744 = J(\tau) = q^{-1} + 196884q + 21493760q^2 + \cdots,$$

where the traditional constant 744 is considered irrelevant.

Then the vector space direct sum of $V(n)$s could be written as:

$$qJ(\tau) = \sum_{n=0} q^n \dim(V_n) = 1 + 196884q^2 + 21493760q^3 + \cdots.$$

Since the irreducible representations of $\mathfrak{M}$ correspond to the traces of its identity element, McKay and Thompson suggested that traces of other elements of $\mathfrak{M}$ should show up as sums in expansions of other j-like functions. These are called mini-j functions (or more formally Hauptmoduls).

In 1979, Conway and Norton tested this idea by checking the character table of $\mathfrak{M}$, which had already been published in the *Atlas of Finite Groups* [Conway *et al.*, 1985]. To understand this we need to look at a couple of character tables.

The 60-element Icosahedral group (also called the Alternating-5 group $\mathcal{A}_5$) has five irreducible representations. Thus the entire Icosahedral character table can be displayed here, where the columns correspond to the five irreducible representations and the rows correspond to the five conjugacy classes.

Icosahedral group character table (five classes)

$1A$	$2A$	$3A$	$5A$	$5B$
1	1	1	1	1
3	-1	0	$\frac{1}{2}(1+\sqrt{5})$	$\frac{1}{2}(1-\sqrt{5})$
3	-1	0	$\frac{1}{2}(1-\sqrt{5})$	$\frac{1}{2}(1+\sqrt{5})$
4	0	1	-1	-1
5	1	-1	0	0

Note that, according to finite group representation theory, the number of inequivalent irreducible representations is always equal to the number of conjugacy classes. Thus any character table is a square matrix consisting of traces of group element matrices. Cf. the Octahedral character table in Chap. 2.

The alternating group $\mathcal{A}_n$, where $n \geq 5$, is the beginning of the series of the Finite Simple Group Classification project. And we note that $\mathcal{A}_5$ as the first simple group in this series was the basis of Galois' proof (1831) that five-degree polynomials have no general solution cf. [Birkhoff and MacLane, 1965].

In contrast to the Icosahedral group ($\mathcal{A}_5$) table, the Monster group $\mathfrak{M}$ table (contained in the *Atlas*) has 194 irreducible representations, and thus 194 conjugacy classes. We can display here only a few of its trace entries.

Also in contrast to the 60 elements of $\mathcal{A}_5$, the Monster group order $|\mathfrak{M}|$ is approximately 8×10^{53}, which was determined in 1973 by Conway as a product of its prime factors [Ronan, 2006]:

$$2^{46} \cdot 3^{20} \cdot 5^9 \cdot 7^6 \cdot 11^2 \cdot 13^2 \cdot 17 \cdot 19 \cdot 23 \cdot 29 \cdot 31 \cdot 41 \cdot 47 \cdot 59 \cdot 71.$$

Monster group character table (194 classes).

$1A$	$2A$	$2B$	$3A$	$\ldots 119B$
1	1	1	1	1
196883	4371	275	...	...
21296876	91884	−2324	...	...
...	...	...	...	...
258823477531055064045234375	...	...	...	...

Conway and Norton calculated that the mini-j function of the $2A$ column was:

$$1/q + c_1 + 4372q + 96256q^2 + 1240002q^3 + \cdots,$$

where $4372 = 1 + 4371$ and $96256 = 1 + 4371 + 91884$ and further such sums of $2A$ traces.

Moreover, the mini-j function of the $2B$ column was:

$$1/q + c_2 + 276q - 2048q^2 + 11202q^3 - \cdots,$$

where $276 = 1 + 275$ and $-2048 = 1 + 275 - 2324$ and further such sums of $2B$ traces.

On the basis of these calculations, Conway and Norton published the "Monstrous Moonshine" conjectures, including especially the claim that all such mini-j functions (Hauptmodul expansions) would correspond to sums of traces from the columns of the Monster group character table. This implies that each conjugacy class of the Monster would correspond to such a Hauptmodul [Conway and Norton, 1979; Conway, 1980].

The most important item in the series of Moonshine conjectures was the claim that the mini-j functions would be Hauptmoduls for genus-0 subgroups Γ of the Modular group $\mathcal{SL}(2, \mathcal{Z})$. Any Hauptmodul is an isomorphism from $\mathcal{H}/\Gamma$ to $\mathbb{C}$, which maps $\mathcal{H}$ to a Riemann surface.

Note that $\mathcal{H}$ is the upper-half part of $\mathbb{C}$ and is topologically equivalent to the hyperbolic plane. Only if $\mathcal{H}/\Gamma$ is mapped to a Riemann sphere $(\mathcal{C} \cup \infty)$ is this Hauptmodul a genus 0 Hauptmodul. Otherwise the mapping would correspond to a Torus (genus 1), or some higher genus surface.

Since the Moonshine conjectures demanded such a genus-0 Hauptmodul for each conjugacy class of $\mathfrak{M}$, one might expect 194 different such Hauptmoduls. However, by calculation, there are only 171 distinctly different genus-0 Hauptmoduls arising from the structure of $\mathfrak{M}$. Moreover, there are 616 genus-0 Houptmoduls, but only 171 of them correspond to the characters of $\mathfrak{M}$ [Gannon, 2006].

We note here the curious fact that in 1975 Andrew Ogg had already calculated that subgroups $\Gamma_0(p)$ of $\mathcal{SL}(2, \mathcal{Z})$ correspond to genus 0 Hauptmoduls only if they are generated by the 2×2 matrices:

$$\begin{pmatrix} 0 & 1 \\ -p & 0 \end{pmatrix}$$

where p is one of the 15 prime numbers:

$$\{2, 3, 5, 7, 11, 13, 17, 19, 23, 29, 31, 41, 47, 59, 71\}.$$

When Ogg presented this result at a conference in Paris, he learned from Jacques Tits that these 15 primes were exactly the 15 prime factors of the Monster group (which had yet to be proven to exist). Ogg offered a bottle of Jack Daniels' whiskey to anyone who could explain this astounding coincidence. Thus Monstrous Moonshine had its beginning [Ronan, 2006].

As was later discovered, for each of these 15 Monster primes p, there is an element g of $\mathcal{M}$ of order p, such that $\Gamma_0(p)$ corresponds to a Monster group Hauptmodul. For example:

$$J_{13}(\tau) = q^{-1} + q + 2q^2 + q^3 + 2q^4 - 2q^5 - 2q^7 - 2q^8 + q^9 + \cdots .$$

Thus while the relationship between $\mathfrak{M}$ and the 15 primes is somewhat clarified, it is still rather mysterious — like many other aspects of Monstrous Moonshine [Gannon, 2006].

In 1980, Robert Griess proved the existence of the Monster group $\mathfrak{M}$ by constructing its 196883D representation. He did this without computer calculations by describing $\mathfrak{M}$ as an automorphism group of a 196883-dim. algebra (which is commutative but non-associative), and is now called the Griess algebra. He also confirmed the correctness of the character table of $\mathfrak{M}$, as already published in the *Atlas* [Griess, 1981, 1982].

In 1983 Frenkel, Lepowsky and Meurman presented a vertex algebra construction of $\mathfrak{M}$ at a conference in Berkeley [Frenkel *et al.*, 1985]. This was a widely ranging paper which entailed the ADE Lie algebras, lattices, and error-correcting codes. These topics will be detailed in future chapters. The important point here is that the vertex algebra construction was borrowed from the mathematics of string theory.

They called their infinite dimensional vertex algebra V-natural, (symbolized as $V\natural$) indicating that it is a natural representation structure for the Monster group. To create $V\natural$, they used the bosonic string geometry. Bosonic strings sweep out a 2D worldsheet in a 26D spacetime. Thus there are 24 transverse dimensions in which the worldsheet vibrates. These dimensions correspond to the 24D Leech lattice L_{24}, so that the bosonic string can be compactified on the torus $\mathfrak{R}^{24}/L_{24}$, which is an orbifold.

$V\natural$ as a vertex operator algebra was constructed as the tensor product of the group ring of L_{24} with the space of oscillators representing the bosonic string vibrations in $\mathfrak{R}^{24}$. Since this oscillator construction is an infinite dimensional polynomial ring, $V\natural$ is an infinite dimensional vertex algebra.

Frenkel, Lepowsky, and Meurman called $V\natural$ the Moonshine module, because they were able to show that the Monster group $\mathfrak{M}$ is the automorphism group of $V\natural$. Moreover, they showed that the graded dimensions of $V^\natural$ are expressed by the Fourier expansion of the j-invariant.

Remark 1. Note that the vertices of the vertex algebra correspond to a stringy version of the vertices of the Feynman diagrams of point particle interactions. These stringy vertices occur as strings interact by joining and separating as they sweep out the 2D worldsheet. Because these stringy vertices are actually of finite length rather than being infinitely small points, the perturbation expansion of the string interaction does not succumb to the infinities of the point-particle interactions.

Remark 2. The Leech lattice L_{24} as an even self-dual lattice can be used to generate the 24-bit Golay code, which is an error-correcting code (to be discussed in more detail in Chap. 13). The automorphism group of the Leech lattice is another sporadic simple group discovered by Conway and is called Co_1. Moreover the Golay code can be constructed from three copies of the E_8 error-correcting code, based on the even, self-dual E_8 lattice [Conway and Sloane, 1988].

Remark 3. E_8 is a 248D Lie algebra. Moreover, because the E_8 lattice is self-dual, both the E_8 fundamental and adjoint representations are of dimensionality 248. It is thus perhaps significant that the constant 744 in the j-invariant is 3(248), which is analogous to the 3(8)D structure of the Leech lattice (and Golay code) as mentioned above. Ordinarily, of course, the constant 744 is simply deleted from the j-function as of no significance. Here I conjecture that 744 is quite significant.

The vertex algebra $V^\natural$ of [Frenkel *et al.*, 1985] as a Moonshine module (based in part on the Leech lattice L_{24}) was still far from proving the Monstrous Moonshine conjectures. These conjectures, most notably, claimed that Hauptmoduls (the mini-j functions) could be assigned to each conjugacy class of $\mathfrak{M}$. The j-function corresponded only to the "trivial" conjugacy class consisting of the identity element of $\mathfrak{M}$, and therefore to sums of irreducible representations of $\mathfrak{M}$. In this sense, the j-function could be considered the Hauptmodul for the entire modular group $\mathcal{SL}(2, \mathcal{Z})$, whereas the (mini-$j$) Hauptmoduls would correspond to subgroups of $\mathcal{SL}(2, \mathcal{Z})$.

These Hauptmodul conjectures were finally proved in 1992 by Richard Borcherds (who had been a graduate student of John Conway at Cambridge University). A key step in this proof was his construction of a Monster Lie algebra. This construction was inspired by the earlier work on the connection between the affine Lie algebra $\mathrm{E}_8^{(1)}$ and $j^{1/3}$, where instead of $j = 1728(G_2)^3/\Delta$ we have $j^{1/3} = 1728(G_2)/\Delta$.

In 1980 Kac and Lepowsky published papers showing that $j^{1/3}$ is the normalized character of the affine $\mathrm{E}_8^{(1)}$ (Kac–Moody) Lie algebra.

Note that the affine Weyl group of any affine Lie algebra $\mathfrak{g}^{(1)}$ is the semi-direct product of the finite Weyl group $\mathfrak{W}(\mathfrak{g})$ with the lattice $\mathcal{Z}^r$ (where r is the rank of $\mathfrak{g}$). Thus the Weyl group of $\mathrm{E}_8^{(1)}$ is $\mathfrak{W}(\mathrm{E}_8)$ s.d. $\otimes \mathcal{Z}^8$ [Kac, 1980; Lepowsky, 1980].

Note also that the ADE Weyl groups are the same as the ADE Coxeter groups as described in Chap. 6.

The Monster Lie algebra $\mathfrak{M}_\mathrm{L}$ would have to be a generalization of an affine Kac–Moody Lie algebra $\mathfrak{g}^{(1)}$. To create $\mathcal{M}_L$ Borcherds used the full structure of the 26D Lorentz spacetime $\mathcal{L}^{25,1}$ of bosonic string theory. The uniqueness of $\mathcal{L}^{25,1}$ can be seen in the purely number theoretic fact that the only sum of square integers which equals a square integer is:

$$0^2 + 1^2 + 2^2 + 3^2 + \cdots + 24^2 = 70^2.$$

Thus for the 26D Lorenz space $\mathcal{L}^{25,1}$ one can write:

$$0^2 + 1^2 + 2^2 + 3^2 + \cdots + 24^2 - 70^2 = 0.$$

so that in $\mathcal{L}^{25,1}$ we have the norm-0 vector with components:

$$(0, 1, 2, \ldots, 24, -70).$$

Such a norm-0 vector is a light-like vector in $\mathcal{L}^{25,1}$.

The even unimodular lattice in $\mathcal{L}^{25,1}$ called $\mathrm{II}_{25,1}$ has as its Weyl vector: $w = (0, 1, 2, 3, \ldots, 24, 70)$. Moreover, transverse to w is the Leech lattice L_{24}. There are actually 23 other 24D lattices which are transversals derived from other light-like vectors in $\mathcal{L}^{25,1}$. These other 24D lattices are called Niemeier lattices, and they are classified by direct sums of ADE Coxeter graphs (and lattices). Thus the corresponding ADE ranks must sum to 24. Moreover, these 23 lattices (like $\mathcal{L}_{24}$) are all even unimodular lattices [Conway and Sloane, 1988].

Neimeier lattice	Coxeter number K	Leech lattice points v
D_{24}	46	25
$D_{16}E_8$	30	26
E_8^3	30	27
A_{24}	25	25
D_{12}^2	22	26
$A_{17}E_7$	18	26
$D_{10}E_7^2$	18	27
$A_{15}D_9$	16	26
D_8^3	14	27
A_{12}^2	13	26
$A_{11}D_7E_6$	12	27
E_6^4	12	28
$A_9^2D_6$	10	27
D_6^4	10	28
A_8^3	9	27
$A_7^2D_5^2$	8	28
A_6^4	7	28
$A_5^4D_4$	6	29
D_4^6	6	30
A_4^6	5	30
A_3^8	4	32
A_2^{12}	3	36
A_1^{24}	2	48

Here the Coxeter number K is listed for the key Coxeter graph in the combination of graphs. For any ADE graph of rank r, hr is the number of roots, which is also the number of spheres that pack around a central sphere in the corresponding Coxeter lattice.

In this 26D context, each 24D Niemeier lattice has $rh = 24h$ vectors of norm 2 (which is the minimal norm). Moreover, each Niemeier lattice corresponds to a deep hole in the Leech lattice; and v is the number of Leech lattice points around the deep hole.

Clearly all these lattice structures are closely related to each other within the overarching lattice structure of $II_{25,1}$.

The specialness of the Leech lattice L_{24} corresponds to the specialness of the light-like vector specified by the norm-0 vector:

$$(0, 1, 2, \dots, 24, -70)$$

which is based on the unique hyper-Pythagorean relationship

$$1^2 + 2^2 + 3^2 + \cdots + 24^2 = 70^2.$$

This was first conjectured by E. Lucas in 1875, as the solution to the "cannonball stacking" problem. If we start with a square array of 24×24 cannonballs and keep stacking cannonballs in a four sided pyramid structure we will end with 1 cannonball at the top and a total of 4900 cannonballs. Following the partial proof of the Lucas conjecture in [Watson, 1918], the final proof was provided by [Anglin, 1990]. Moreover, David Morrison, as in [Baez, 2002], pointed out that:

$$1^2 + 2^2 + 3^2 + \cdots + n^2 = m^2 = n(n+1)(2n+1)/6.$$

Therefore n and m can be determined by finding all the rational points (n, m) on the elliptic curve:

$$(1/3)n^3 + (1/2)n^2 + (1/6)n = m^2$$

which has the solution $n = 24, m = 70$ as the only integer solution.

Thus we could think of this elliptic curve as the Monster elliptic curve. This is because the Monster Lie algebra (as constructed by Borcherds) has as its Weyl vector:

$$w = (0, 1, 2, 3, \dots, 24; 70)$$

so that its simple roots can be defined as vectors r in $II_{25,1}$ such that:

$$r \cdot r = 2, \quad r \cdot w = -1.$$

There is an infinity of such simple roots and they are identical to the points of the Leech lattice L_{24}.

The automorphism group $\mathrm{Aut}(II_{25,1})$ has as its reflection subgroup the Coxeter group defined by using L_{24} as a Coxeter graph of infinite rank. Thus:

$$\mathrm{Aut}(II_{25,1}) = \mathrm{Cox}(II_{25,1}) \times \mathrm{Per}(L_{24}) \times (1, -1),$$

where $\mathrm{Per}(L_{24})$ is the graph permutation group of L_{24} as a Coxeter graph. Note that while the Conway group Co_1 is the automorphism group of L_{24}, here $\mathrm{Per}(L_{24})$ is equivalent to $\mathrm{Co}_1 \times$ tranlations of L_{24}.

To construct his Monster Lie algebra, Borcherds made use of the "no ghost" theorem from bosonic string theory [Goddard and Thorn, 1972].

The light-like (norm 0) vectors in $\mathcal{L}^{25,1}$ can be used to define the light-cone gauge in 26D bosonic string theory. This has the effect of suppressing all longitudinal vibrations in the 2D worldsheet. The vibrational states are thus confined to the 24D transverse coordinates where the 24D Leech lattice L_{24} can be constructed.

Note that L_{24} has no vectors of minimal norm 2, but has norms for every even number beyond 2. This is in contrast to the 23 Niemeier lattices, which have vectors of minimal norm 2 [Conway and Sloane, 1988].

There are 196,560 norm-4 points; and this corresponds to the 196,560 unit spheres packing perfectly around a central sphere. (For more on sphere-packing mathematics, see Chap. 13.)

The consequence of the L_{24} description of the transverse states in $\mathcal{L}^{25,1}$ is that there are no so-called "ghost states" (negative-norm states) as physical states.

The disadvantage of the light-cone gauge fixing is that while unitarity is preserved, Lorentz invariance becomes difficult to prove. However, for 26D spacetime Lorentz invariance is provable. Indeed from Lorentz invariance and a zero-point calculation on the bosonic strings, one can derive the 26D bosonic spacetime [Brink and Nielsen, 1973; Becker, Becker, and Schwarz, 2007; Polchinski Vol. 1, 1998].

Borcherds constructed the Monster Lie algebra as a generalized Kac–Moody algebra. This makes use of the 2D even Lorentzian lattice $\mathrm{II}_{1,1}$, which can be considered as a sublattice of the 26D even unimodular lattice $\mathrm{II}_{25,1}$. This 2D Lorentzian lattice corresponds to the 2D Cartan subalgebra of the Monster Lie algebra. The Dynkin diagram would be:

$$\mathrm{O{=}O},$$

where the double lines indicate a 135° angle between the basis roots. There is one simple real root $(1, -1)$, and an infinity of imaginary simple roots $(1, n)$, where $n > 0$ is an integer.

Moreover, these imaginary simple roots generate other roots of multiplicity $c(n)$, where c is the dimension of the root space $V(n)$ as a subspace of the infinite vector space V. Rather amazingly, these multiplicities $c(n)$ happen to be the coefficients of the j-function, so that V is the vertex algebra $V\natural$, which is the Monster module of [Frenkel *et al.*, 1985].

Since the coefficients of the j-function correspond to sums of irreducible representations of the Monster group $\mathfrak{M}$, we are still accounting only for characters of the identity element (i.e. the first column of the character table).

However, Monstrous Moonshine claims that all the characters of $\mathfrak{M}$ can be accounted for by mini-j functions (Hautpmoduls).

To prove this Monstrous Moonshine conjecture, Borcherds had to account for the McKay–Thompson functions $T_g(\tau)$, where g is any element of $\mathfrak{M}$. His first step was to show that $T_g(\tau)$ are completely replicable functions (i.e. that their coefficients obey certain identities). For example, $T_1(\tau)$ as the elliptical modular function

$$j(\tau) - 744 = \sum c(n)q^n$$

satisfies the "twisted" Weyl denominator formula:

$$j(\sigma) - j(\tau) = p^{-1}\Pi(1 - p^m p^n)^{c(mn)},$$

where $m > 0$ and $n \in Z; p = e^{2\pi i\sigma}$ and $q = e^{2\pi i\tau}$.

Moreover, Borcherds was able to construct analogous twisted denominator formulas for each character of $\mathfrak{M}$ (and thus for each element g). These denominator formulas are the analogs of the denominator formula (above) for the identity element.

These denominator formulas have recursion relations with the coefficients of the McKay–Thompson functions $T_g(\tau)$. The Conway–Norton Moonshine functions agree with these recursion relations through the first seven terms, which was enough to convince Borcherds (and other mathematicians) that Monstrous Moonshine is valid [Borcherds, 1992].

For this and much related work, Borcherds received the Fields Medal in 1998.

There are similar Moonshine conjectures for other sporadic simple groups such as the Baby Monster. So it is useful at this point to list all 26 sporadic simple groups, and some of their relations to the

Monster group. Here the approximate group order $\sim 10^x$ is listed as (x) appended to the sporadic group label.

Mathieu groups: $M_{11}(4), M_{12}(5), M_{22}(7), M_{23}(8), M_{24}(9)$

Janko groups: $J_1(5), J_2(6), J_3(8), J_4(20)$

Conway groups: $\mathrm{Co}_1(18)$, $\mathrm{Co}_2(13)$, $\mathrm{Co}_3(11)$

Fischer groups: $\mathrm{Fi}_{22}(14)$, $\mathrm{Fi}_{23}(18)$, $\mathrm{Fi}_{24}(24)$

Higman–Sims group: HS(7)

McLaughlin group: McL(9)

Held group: He(9)

Rudvalis group: Ru(11)

Suzuki sporadic group: Suz(11)

O'Nan group: O'N(11)

Harada–Norton group: HN(14)

Lyons group: Ly(16)

Thompson group: Th(17)

Baby Monster group: B(33)

Fischer–Griess Monster group: $\mathfrak{M}(54)$

Nineteen of these groups are subgroups or quotient groups of $\mathfrak{M}$. This leaves six groups as not involved in M_{24} so they are called the pariah groups:

J_1, J_3, J_4, O'N, Ru, and Ly (J_1 being a subgroup of O'N).

Note, however, that the pariah J_4 has M_{24} as a subgroup, while M_{24} is a subgroup of Co_1, which is a subgroup of $\mathfrak{M}$. Also the pariah Ly has McL as a subgroup, while McL is a subgroup of Co_2 (which is a subgroup of Co_1). Moreover McL has both M_{11} and M_{12} as subgroups.

There are many more such subgroup relations in the complex web of subgroup relationships within $\mathfrak{M}$ [Conway, 1980; Conway and Sloane, 1988; Ronan, 2006].

The original Monstrous Moonshine paper [Conway and Norton, 1979] proposed also a generalized Moonshine for other sporadic groups. Subsequently Larissa Queen (and others) were able to fulfill this prediction by constructing Hauptmoduls for the Conway groups (Co_1, Co_2, Co_3) [Queen, 1981]. This entailed the use of the Dedekind eta function, in the relationship between the j-function and the McKay–Thompson series for the Conway groups. For example:

$$\begin{aligned} j_{4A}(\tau) &= T_{4A}(\tau) + 24 = [\eta^2(2\tau)/\eta(\tau)\eta(4\tau)]^{24} \\ &= 1/q + 24 + 276q + 2048q^2 + 11202q^3 + 49152q^4 + \cdots, \end{aligned}$$

where $\eta(\tau)$ is the Dedekind eta function [MacDonald, 1972; Kac, 1978].

The sporadic simple groups were the final result of the vast enterprise to classify all simple finite groups. The 26 sporadic simple groups are those finite simple groups that do not fit into the four infinite series of such groups [Gorenstein, 1982].

Thus we can note an analogy between the classification of the ADE Coxeter groups (and Lie algebras) and the classification of the simple finite groups.

The ADE classification entails two infinite series (A and D) and an exceptional series (E_6, E_7, E_8).

Similarly the finite simple group classification entails two infinite series and an exceptional series:

The Alternating group series: $\mathcal{A}_n (n \geq 5)$

The Groups of Lie type (Chevalley groups)

The 26 sporadic groups (27 if the Tits group is included).

(**Note:** the cyclic groups of prime order are also simple groups, while the cyclic groups of any order are the McKay groups of type A_n.)

There are already sphere-packing analogies between $\mathfrak{M}$ and E_8. Presumably, there are many other analogies between the ADE classification scheme and the finite simple group classification scheme.

Chapter 13

Sphere Packing and Error-Correcting Codes

As we have seen in Chap. 12, the 24D Leech lattice L_{24} has played a fundamental role in both the Monster group and string theory, which was used in Borcherd's proof of Monstrous Moonshine.

Here we will introduce another fundamental aspect of the Leech lattice L_{24}: it provides the closest possible packing of spheres in 24D Euclidean space. Indeed, the 196,560 unit spheres which pack around any unit sphere in L_{24} can be used to generate the Golay-24 code, $\mathcal{G}_{24}$. This is an error-correcting code, in which each codeword consists of 24 binary digits: 12 message digits and 12 error-correcting digits. The Golay-24 code is a very powerful code since it corrects for any 3 errors occurring anywhere in the 24-digit word [Conway and Sloane, 1988].

In general, error-correcting code theory is a concrete implementation of Claude Shannon's information theory. Shannon defined information by way of the entropy function:

$$H = \sum p_i \log 1/p_i,$$

where Σ is the sum over an alphabet of n symbols, each of probability p_i. Usually the logarithm is assumed to be base 2, which corresponds to binary digits. So this implies that the information using an alphabet of n symbols is encoded into a codewords consisting of binary digits.

Shannon's information theory provides bounds for any such system transmitting information through a noisy channel [Shannon, 1948, 1949].

Coding theorists provided the actual error-correcting codes to implement the abstract ideas of Shannon's coding theory [Hamming, 1980].

Note also that error-correcting codes are essential for keeping stored information (such as in computers) error free [McEliece, 1985].

It is significant that Shannon was working closely with Richard Hamming at Bell labs when he published his information theory. Indeed, Shannon describes the Hamming-7 code $\mathcal{H}_7$ (with 4 message carrying digits, and 3 error-correcting digits) as a concrete implementation of his theory.

Moreover, Golay who had previously worked at Bell labs published his Golay codes $(\mathcal{G}_{11}, \mathcal{G}_{12}, \mathcal{G}_{23}, \mathcal{G}_{24})$ while he was working at the U.S. Army Signal Corp [Golay, 1949, 1954].

There are, of course, a vast number of error-correction codes [Berlekamp, 1968; MacWilliams and Sloane, 1977]. Here, however, we will concentrate on the fundamental codes based on the ADE Coxeter lattices as emphasized by [Conway and Sloane, 1988]. This will keep us closer to applications in physics.

Of special importance for both coding theory and physics are the self-dual codes. This is because [Gleason, 1971] proved that the weight enumerators of self-dual codes are invariant polynomials. Some of these polynomials are the ones encountered in Chaps. 6–8.

Moreover, Gleason proved that the weight enumerator for any even self-dual binary code is an element of the polynomial algebra $\mathcal{C}[a, b]$, where a and b are certain invariant polynomials. (Here "even" means that the distance between any two codewords is a multiple of 4.)

a is the Hamming-8 weight enumerator: $h = x^8 + 14x^4y^4 + y^8$
b is the Golay-24 weight enumerator: $g' = x^4y^4(x^4 - y^4)^4$.

Note that g' is used as a more convenient generator of $\mathbb{C}[a, b]$ than the actual Golay-24 weight enumerator which is:

$$g = x^{24} + 759x^{16}y^8 + 2576x^{12}y^{12} + 759x^8y^{16} + y^{24}$$

so that $g' = (h^3 - g)/42 = x^4y^4(x^4 - y^4)^4$.

This can be rewritten in terms of the relative octahedral invariants, f and h, which are also absolute tetrahedral invariants (cf. Chap. 7).

$$f = x^5 y - y x^5,$$
$$h = x^8 + 14x^4y^4 + y^8.$$

Then, since $f = xy(x^4 - y^4)$,

$$f^4 = x^4y^4(x^4 - y^4)^4 = g'.$$

So we can write: $f^4 = (h^3 - g)/42$,

$$g = h^3 - 42f^4.$$

Curiously, this is analogous to the tetrahedral syzygy formula:

$$j^2 = h^3 - 108f^4,$$

where j is the absolute tetrahedral invariant:

$$j = x^{12} - 33x^8y^4 - 33x^4y^8 + y^{12}$$

and is thus analogous to the octahedral syzygy formula:

$$f^2j^2 = f^2h^3 - 108f^6$$
$$Z^2 = XY^3 - 108X^3,$$

where X, Y and Z are the absolute octahedral invariants.

The zero sets of these syzygies correspond to the E_6 and E_7 gravitational instanton structures of $\mathbb{C}^2/\mathcal{TD}$ and $\mathbb{C}^2/\mathcal{OD}$, respectively (cf. Chap. 8).

Here E_8 is also involved, because h is the Hamming-8 weight enumerator, and the Hamming-8 code can be constructed from the E_8 lattice, which is an even self-dual lattice. Moreover, the $E_8 \otimes E_8$ lattice is a sub-lattice of the Leech lattice L_{24}, which then provides the structure for the Golay-24 code. This is a very powerful code that corrects any of 3 errors anywhere within the 24-bit codewords. It will also detect as many as 7 errors within these codewords [Peng, 2006].

According to Gleason's theorem, these even self-dual codes are called Type **II** codes, which fit into the three types of all self-dual (binary and ternary) codes. We can list the invariant polynomials for the polynomial algebra $\mathbb{C}[\mathcal{a}, \mathcal{b}]$ for Types **I** and **III** as follows:

Type **I**: (binary): $\mathcal{a} = i_2 = x^2 + y^2; \quad \mathcal{b} = \mathcal{h}' = x^2y^2(x^2 - y^2)^2$

Type **III**: (ternary): $\mathcal{a} = t_4 = x^4 + 8xy^3; \quad \mathcal{b} = t\mathcal{g}_{12} = (y^3(x^3 - y^3)^3,$

where t_4 is the weight enumerator for the tetracode; and $t\mathcal{g}_{12}$ is the weight enumerator for the ternary Golay code.

Another example of a ternary code is generated by the E_6 lattice [Conway and Sloane, 1988; MacWilliams and Sloane, 1977; Sloane, 2006].

Remark. Andrew Gleason (1921–2008) is more famous among physicists for "Gleason's Theorem" (1957), which was a precursor to John Bell's theorem (1964) on quantum entanglement. Among coding theorists, "Gleason's Theorem" (1971) is the one described above cf. [Gleason, 1957; 1971; Bell, 1987; Bolker, 2009].

It is significant that the quantum entanglement of qubits plays a fundamental role in quantum information theory. Accordingly, quantum error correction codes will be necessary to ensure computational reliability. Thus we conjecture that there must be some intimate relationship between the two quite different "Gleason's Theorems."

Having introduced these fundamental error-correcting code abstractions, it is now necessary to describe the more concrete aspects of these codes.

In general, a weight enumerator (WE) for a binary code is an abstract polynomial formula that corresponds to the distribution of 1's and 0's in the set of codewords.

For example, in the case of the Hamming-8 code, the WE is:

$$\mathcal{h} = x^8 + 14x^4y^4 + y^8.$$

This means that there are 16(=1 + 14 + 1) codewords:

1 consisting of 8 zeros, 1 consisting of 8 ones, and

14 consisting of 4 ones and 4 zeros.

We can list the 16 Hamming-8 codewords (with 4 message carrying digits and 4 error correcting digits) as follows:

0: 0000 0000	8: 1111 1111
1: 1011 1000	9: 0100 0111
2: 0101 1100	10: 1010 0011
3: 0010 1110	11: 1101 0001
4: 1001 0110	12: 0110 1001
5: 1100 1010	13: 0011 0101
6: 1110 0100	14: 0001 1011
7: 0111 0010	15: 1000 1101

Notice that the set of codewords numbered 8 through 15 are complements of the codewords numbered 0 through 7.

There are, of course, 2^8 (=256) possible 8-bit strings. However, those 8-bit strings which contain at most 1 error (a flip between 0 and 1) will be read correctly by this set of codewords. Moreover, this Hamming-8 code will also detect a possible second error in the 8-bit string.

It is also noteworthy that the Hamming-7 code $\mathcal{H}_7$ can be derived from the $\mathcal{H}_8$ code by removing the last digit from each codeword. Moreover, the $\mathcal{H}_7$ code corrects for 1 error anywhere in the 7-bit string. This code is an example of a perfect code in the sense that it uses its coding space most efficiently.

However, because the $\mathcal{H}_7$ code is not a self-dual code, there exists a dual code $\mathcal{H}_7^{\perp}$ with 3 message carrying digits and 4 error-correcting digits. This reversal from a [7,4] code to a [7,3] code corresponds to deriving the $\mathcal{H}_7^{\perp}$ codewords from the $\mathcal{H}_7$ codewords. There are $2^3 = 8$ codewords; and the 3 message carrying digits are exactly the error-correcting digits of the $\mathcal{H}_7$ codewords, while the 4 error correcting digits of $\mathcal{H}_7^{\perp}$ are exactly the message digits of the $\mathcal{H}_7$ codewords. We can list these 8 $\mathcal{H}_7^{\perp}$ codewords (derived from $\mathcal{H}_7$) as:

0: 000 0000	4: 011 1001
1: 100 1011	5: 101 1100
2: 110 0101	6: 010 1110
3: 111 0010	7: 001 0111

Notice that (aside from the 0 codeword) the digits cycle through the codewords. This is the cyclic code $\mathcal{S}_3$ of type $[2^3-1,3]$. It is also a Simplex code as the dual of the Hamming code $\mathcal{H}_8$, which implies that the codewords form an 8 vertex simplex in 7D Euclidean space [Sloane, 1979; 2006; Mac Williams and Sloane, 1977].

It is significant that $\mathcal{H}_7$ corresponds to the E_7 weight lattice, while $\mathcal{H}_7^\perp$ corresponds to the E_7 root lattice. In fact the 7 simple positive roots correspond the codewords 1 through 7, while the zero root corresponds codeword 0. For details on weight and root lattices see Chap. 5.

Since the root lattice is a sub-lattice of the weight lattice, it is not surprising that the 8 $\mathcal{H}_7^\perp$ codewords comprise a substructure within the 16 $\mathcal{H}_7$ codewords.

Of greatest importance for physics is the fact that $\mathcal{H}_8$ can be generated by the E_8 lattice of the E_8 Lie algebra. In string theory the E_8 lattice plays a key role. This is because the $E_8 \otimes E_8$ lattice (which is self-dual just as is the E_8 lattice) provides anomaly cancellation for the heterotic string theory. This theory is called heterotic because it is a subtle weaving together of the 10D spacetime of superstring theory and the 26D spacetime of the bosonic string theory. This is a closed string theory, in which the two chiralities of the string see different dimensions of spacetime. The left movers (rotating counter-clockwise) see the full 26 bosonic dimensions; while the right movers (rotating clockwise) see only the ten supersymmetic dimensions.

Moreover, the 16D $E_8 \otimes E_8$ lattice interpolates the dimensional mismatch between the 10D spacetime and the 26D spacetime. In fact, for the 24 dimensions transverse to the 2D worldsheet, we really need the 24D Leech lattice L_{24} which has $E_8 \otimes E_8$ as a sub-lattice [Conway and Sloane, 1988].

As described in Chap. 12, L_{24} provides the structure to model the 24 transverse dimensions of the 26D bosonic string theory. Then by compactifying 16 of these dimensions by the 16D torus:

$$\mathfrak{R}^{16}/L_{16},$$

where L_{16} is the $E_8 \otimes E_8$ lattice, we can have a consistent superstring theory defined on the other ten dimensions of the 26D heterotic theory.

There is an alternative heterotic theory, in which L_{16} is the $\mathrm{D}_{16}/\mathcal{Z}_2$ lattice, and we note that D_{16} is the Lie algebra whose compact Lie group is $\mathcal{SO}(32)$.

In both versions of the heterotic theory the Lie group generated by the lattice is the gauge group of the superstring theory.

Note that the self-duality of $\mathrm{E}_8 \otimes \mathrm{E}_8$ (or $\mathrm{D}_{16}/\mathcal{Z}_2$) is necessary for anomaly cancellation. This self-duality is implicit in the fact that the E_8 root lattice is identical to the E_8 weight lattice (and likewise for $\mathrm{D}_{16}/\mathcal{Z}_2$). In fundamental particle theory, the root lattice carries bosonic eigenvalues, while the weight lattice carries fermionic eigenvalues. So in the heterotic theory the main distinction between bosons and fermions is the dimensionalities of the relevant spacetimes: 26 and 10.

Remark. It is perhaps surprising to mathematicians (and to physicists) that basic mathematical distinctions correspond so precisely to basic particle physics distinctions (cf. Chap. 5):

Root lattice (adjoint rep.) $\Leftrightarrow$ Bosons (integral spin)

Principal fiber bundle $\Leftrightarrow$ Gauge particle eigenvalues

Weight lattice (fundamental rep.) $\Leftrightarrow$ Fermions (1/2 integral spin)

Associate vector bundle $\Leftrightarrow$ Matter particle eigenvalues

In this chapter we are emphasizing another such correspondence:

Error-correcting code $\Leftrightarrow$ Anomaly cancellation

Since E_8 is self-dual, its weight lattice is the same as its root lattice. Thus the fundamental representation is the same as its adjoint representation, which is 248D.

The $\mathrm{E}_8 \otimes \mathrm{E}_8$ Heterotic symmetry structure entails a 496D representation. It has been noticed that 496 is a perfect number, the third in the infinite series beginning with 6, 28, 496. As defined by the Greeks, a perfect number is the sum of its divisors; and as discovered by modern mathematicians, there is a 1–1 correspondence between perfect numbers and Mersenne primes [Beiler, 1966].

Note that the alternative heterotic theory has the symmetry structure provided by D_{16} (with compact Lie group $\mathcal{SO}(32)$), so that there is a duality with Type **I** string theory, which has symmetry group $\mathcal{SO}(32)$. Both D_{16} and $\mathcal{SO}(32)$ have 496 as their total dimensions.

Thus, for the series of certain Lie algebras, we have the progression of perfect number dimensions, which correspond to Mersenne primes:

n	Mersenne prime: $2^n - 1$	Perfect number: $(2^{n-1})(2^n - 1)$
2	3	$6 - A_1 \oplus A_1 - \mathcal{SU}(2) \otimes \mathcal{SU}(2) = \mathcal{SO}(4)$
3	7	$28 - D_4 - \mathcal{SO}(8)$
5	31	$496 - D_{16} - \mathcal{SO}(32)$ (& $E_8 \otimes E_8$)
7	127	$8128 - D_{64} - \mathcal{SO}(128)$
13	8191	$33{,}550{,}336 - D_{4096} - \mathcal{SO}(8192)$

Note that the 5th perfect number is 33,550,336, which is the dimension of the D_{4096} Lie algebra, with $\mathcal{SO}(8192)$ as its compact Lie group.

Now 8192 is 2^{13} and is a very interesting number. It is called a Bhaskara twin because twice its square is a cube while twice its cube is a square. Cf. [Sloane: http://oeis.org/A106318]

$$2(8192)^2 = (512)^3$$
$$2(8192)^3 = (1048576)^2.$$

Also we note that the Cartan subalgebra (and thus the lattice) of $\mathcal{D}_{4096}$ is 4096D. Curiously $4096 = 2^{12}$ and is therefore the number of codewords of the Golay-24 code, which has 12 message carrying bits, and 12 error-correcting bits with weight distribution:

$$1 + 759 + 2576 + 759 + 1 = 4096.$$

Of course, since perfect numbers are always triangular numbers and the dimensionality of the D-type Lie algebra is always a triangular number, there will always be some pairing of perfect numbers with certain D-type Lie algebra. The only exception is right at the beginning with $A_1 \oplus A_1$ being a 6D Lie algebra, with $\mathcal{SU}(2) \otimes \mathcal{SU}(2)$ as its compact Lie group. Since $\mathcal{SU}(2) \otimes \mathcal{SU}(2) = \mathcal{SO}(4)$, we see that it is the $\mathcal{SO}(n)$ series that is implicit in the triangular (and perfect) number series.

Note that $\mathcal{SO}(n)$ as a rotation group on $\mathfrak{R}^n$ always has rotational degrees of freedom equal to the total dimension of $\mathcal{SO}(n)$, which is equal to the triangular number:

$$(n^2 - n)/2.$$

This formula also records the number of combinations of n things taken 2 at a time, which can be pictured as the number of planes of rotation in the $\mathfrak{R}^n$ space.

Let us now display the ADE hierarchy of sphere-packing lattices. The Leech lattice L_{24} embeds them all [Conway and Sloane, 1988].

Dimension:	1	2	3	4	5	6	7	8	24
Packings:	Z	A_2	A_3	D_4	D_5	E_6	E_7	E_8	L_{24}
Kissing #:	2	6	12	24	40	72	126	240	196560
Quantizer:	Z	A_2	A_3^*	D_4	D_5^*	E_6^*	E_7^*	E_8	L_{24}

Note: lattice X_n^* is dual to lattice X_n.

The Kissing number is the number of unit spheres that pack around any single unit sphere in the sphere packing. The points of intersection are called Kissing points, and they form a symmetric pattern equivalent to the non-zero roots of the ADE Lie algebras cf. [Conway and Sloane, 1988].

The best quantizers correspond to the weight lattices of the ADE Lie algebras. They will be described more fully in Chap. 14.

It is significant that this hierarchy of sphere packings, from E_8 down to A_2, corresponds to the hierarchy of particle physics symmetry groups in grand unified theory as a substructure in string theory.

Moreover, the Leech lattice L_{24} (which contains this entire hierarchy) also plays a fundamental role in heterotic string theory as described above.

It is perhaps confusing that the E_7 root lattice (which provides the densest packing with 126 spheres packed around a central sphere) corresponds to a 8-codeword subcode $\mathcal{H}_7^{\perp}$ of the 16-codeword $\mathcal{H}_7$ Hamming code, which corresponds to the E_7^* weight lattice.

In general then, for lattices which are not self-dual:

$$X_n^* \Leftrightarrow \mathcal{C}_n,$$
$$X_n \Leftrightarrow \mathcal{C}_n^{\perp}.$$

In heterotic string theory, one contemplates the hierarchy of symmetry structures:

$$\mathrm{E}_8 \oplus \mathrm{E}_8 \to \mathrm{E}_8 \to \mathrm{E}_6 \to \mathrm{D}_5 \to \mathrm{A}_4 \to \mathrm{A}_2 \to \mathrm{A}_1,$$

where $\mathrm{A}_2 \oplus \mathrm{E}_6$ as a subalgebra of E_8 (with maximal subgroup $\mathcal{SU}(3) \otimes \mathrm{E}_6$) is expected to be significant, since $\mathcal{SU}(3)$ could refer to a symmetry group acting on the three fermion families.

Thus it is interesting to look at the coding theory aspect of E_6. This is a rather peculiar case, because the E_6 lattice corresponds to a ternary code (rather than a binary code). This code can be modeled by considering the 6D lattice as embedded in complex-3 space $\mathbb{C}^3$ where each of the three dimensions is a copy of the complex plane $\mathbb{C}$ with a triangular grid pattern like that of A_2 as described in Chap. 4. However, for the E_6 lattice, the triangular grid on $\mathbb{C}$ has Eisenstein integers as grid points. These are complex numbers of the form:

$$z = a + b\omega,$$

where a and b are integers; and $\omega = (-1 + i\sqrt{3})/2 = e^{2\pi i/3}$, a complex cube root of unity.

This Eisenstein lattice in $\mathbb{C}^3$ is equivalent to the E_6 root lattice in $\mathfrak{R}^6$. The error-correcting code implicit in this lattice is the 3-codword repetition code, with mod 3 arithmetic [Conway and Sloane, 1988]:

$$[3, 1, 3]\colon (000), (111), (222).$$

The dual lattice E_6^* is the E_6 weight lattice; and the dual code [3, 2, 2] has $9 = 3^2$ codewords, which are vectors summing to 0 (mod 3) in $\mathbb{C}^3$:

$$(000), (111), (222), (1,2,0), (0,1,2), (2,0,1), (0,2,1), (2,1,0), (1,0,2).$$

Thus we see that the E_6 code [3,1,3] is a subcode of the E_6^* code [3,2,2]. This accords with the fact that the $6 + \underline{6}$ simple roots of E_6 are a substructure within the $27 + \underline{27}$ basic weights of E_6.

There are 72 roots which correspond to the 78 ($= 72 + 6$) adjoint representation of E_6. The 27 weights correspond to the fundamental 27D representation, while the combination of positive and negative basic weights corresponds to a 54D representation of E_6.

These roots and weights form the vertices of 6D polytopes in $\mathfrak{R}^6$, and beautiful projections of these polytopes may be seen in Coxeter's beautiful book, *Regular Complex Polytopes* [Coxeter, 1991]. In Coxeter's book these polytopes are considered as complex polytopes in $\mathbb{C}^3$. This depiction is possible only because vertices are points, whose real and complex form are the same (whereas, for example, a complex line is a 2D real object).

The 72-vertex polytope (in Coxeter's 2D projection) may also be seen in the Wikipedia article, "E_6 mathematics."

A more important tertiary code is the Golay-12 code $\mathcal{G}_{12}$, which is a Type **III** code as described in Gleason's theorem on self-dual codes. It is designated as a [12, 6, 6] code, and has the complete weight enumerator:

$$X^{12} + Y^{12} + Z^{12} + 22(Y^6Z^6 + Z^6X^6 + X^6Y^6) \\ + 220X^3Y^3Z^3(X^3 + Y^3 + Z^3).$$

Thus we have $3 + 66 + 660 = 729$ codewords. Rather than list all 729 codewords we will display the generator matrix for $\mathcal{G}_{12}$.

$$\begin{vmatrix} 1\,0\,0\,0\,0\,0\,0\,1\,1\,1\,1\,1 \\ 0\,1\,0\,0\,0\,0\,2\,0\,1\,2\,2\,1 \\ 0\,0\,1\,0\,0\,0\,2\,1\,0\,1\,2\,2 \\ 0\,0\,0\,1\,0\,0\,2\,2\,1\,0\,1\,2 \\ 0\,0\,0\,0\,1\,0\,2\,2\,2\,1\,0\,1 \\ 0\,0\,0\,0\,0\,1\,2\,1\,2\,2\,1\,0 \end{vmatrix}$$

Every codeword is a linear sum (mod 3) over these vectors. And every codeword sums to 0 (mod 3) [Conway and Sloane, 1988].

The Mathieu-24 group $\mathcal{M}_{24}$ of order 244,823,040 is the automorphism group of the Steiner system S(5,8,24), and also the automorphism group of the Golay code $\mathcal{G}_{24}$.

In fact the five sporadic simple Mathieu groups are automorphism groups of Steiner systems; and four of them are Golay code symmetry groups [MacWilliams and Sloane, 1977; Conway and Sloane, 1988].

Group	Order	Steiner syst.	Code	Type	Codewords
$\mathcal{M}_{11}$	7920	S(4,5,11)	$\mathcal{G}_{11}$	Perfect tertiary	729
$\mathcal{M}_{12}$	95040	S(5,6,12)	$\mathcal{G}_{12}$	Self-dual tertiary	729
$\mathcal{M}_{22}$	443,520	S(3,6,22)			
$\mathcal{M}_{23}$	10,200,960	S(4,7,23)	$\mathcal{G}_{23}$	Perfect binary	4096
$\mathcal{M}_{24}$	244,823,040	S(5,8,24)	$\mathcal{G}_{24}$	Self-dual binary	4096

Note: (more precisely) the automorphism groups of $\mathcal{G}_{11}$ and $\mathcal{G}_{12}$ are $\mathcal{M}_{11}/\mathcal{Z}_2$ and $\mathcal{M}_{12}/\mathcal{Z}_2$, respectively. Also $\mathcal{M}_{22}$ is the index-2 subgroup of the full automorphism group of S(3,6,22).
Note also that perfect Golay codes are derived from self-dual codes by deleting the last digit of each codeword. Thus the number of codewords is unaltered.

The five simple Mathieu groups were discovered by Emile Maathieu beteween 1861 and 1873. He described each $\mathcal{M}_n$ group as a permutation group on n elements. They were the first of the sporadic finite simple groups. By 1980 it was known that there were 21 more of these sporadic groups and that the total number was 26, with the largest being the Monster group, with around 10^{53} elements. This completed the classification of all finite simple groups, a project carried out by many mathematicians from 1965 to 1980. The Monster contains 19 of the sporadic groups as subgroups including the five Mathieu groups [Conway and Sloane, 1988] cf. Chap. 12.

A Steiner system $\mathrm{S}(t, k, v)$ is abstractly defined as a set S of v points, with k subsets (called blocks) of S, such that: any t points are contained in only 1 block.

More informally, we can view $\mathrm{S}(t, k, v)$ as a collection of k-person committees chosen from a pool of v people, such that any t persons serve on only one committee.

More generally, such a set of committees could be defined such that any t persons serve on λ committees. This definition would correspond to the structure of a t-design, with the standard label:

$$t - (v, k, \lambda).$$

A very pretty example is the projective plane of order 2. This is the t-design: 2-(7, 3, 1), which is equivalent to the Steiner system S(2, 3, 7). The 7 blocks of the points labeled $\{1, 2, 3, 4, 5, 6, 7\}$ are:

$$123, 345, 561, 174, 275, 376, 426.$$

This is also called the Fano plane, one depiction of which is:

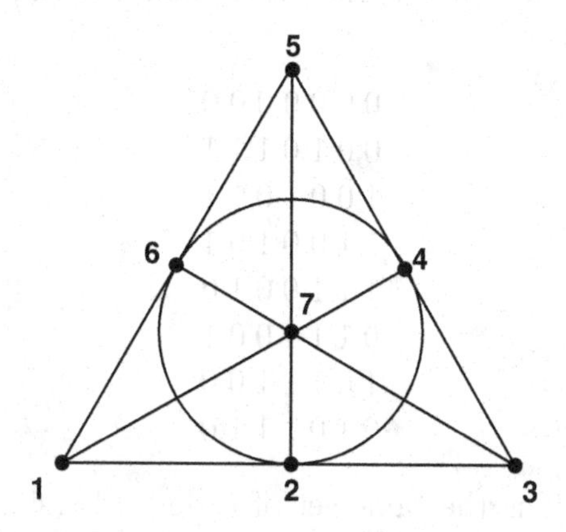

The Fano plane has many applications, only a small number of which can be considered here. There seems to be a special relationship between the Fano plane and the E_7 structures.

For example, as [Baez, 2002] has emphasized, the Fano plane can be used to depict the multiplication structure of the 7 octonions, whose 8th element is the identity. Cf. Chap. 14.

Since the labeling of the 7 points on the Fano plane is arbitrary, we can relabel the points on the above Fano plane depiction. We thus can have 7 triplets of points corresponding to 7 lines:

$$124, 235, 346, 457, 156, 267, 137$$

More abstractly, we can convey this structure in the matrix:

1 1 0 1 0 0 0
0 1 1 0 1 0 0
0 0 1 1 0 1 0
0 0 0 1 1 0 1
1 0 0 0 1 1 0
0 1 0 0 0 1 1
1 0 1 0 0 0 1

where the columns refer to points and the rows refer to lines. For example, the first row corresponds to the line containing points labeled 1 2 and 4 [Sloane, 1981].

This matrix corresponds to the $\mathcal{H}_7^\perp$ code, which can be derived from the E_7 root lattice. The 8 codewords of $\mathcal{H}_7^\perp$ are depicted as complements of the rows of the Fano-plane matrix, along with the zero codeword.

0 0 0 0 0 0 0
0 0 1 0 1 1 1
1 0 0 1 0 1 1
1 1 0 0 1 0 1
1 1 1 0 0 1 0
0 1 1 1 0 0 1
1 0 1 1 1 0 0
0 1 0 1 1 1 0

Note that this is the same set of codewords (slightly reordered) as the $\mathcal{H}_7^\perp$ codewords listed previously.

The automorphism group of $\mathcal{H}_7^\perp$ is Klein's group of order 168 [Conway and Sloane, 1988, p. 80]. This is PSL(2,7), which is the quotient group $\mathcal{SL}(2,7)/\pm 1$, where $\mathcal{SL}(2,7)$ is the set of 2×2 matrices over the finite field F_7 with unit determinant.

This group is called Klein's quartic group because it is the symmetry group of the quartic, cf. [Klein, 1998]:

$$X^3Y + Y^3Z + Z^3X = 0$$

which is a Riemann surface, described as the most symmetrical curve of genus 3 over $\mathbb{C}$.

There are many beautiful depictions of the tiling of this hyperbolic surface with 168 white triangles and 168 black triangles. See, for example [Rauch and Lewittes, 1970; Gray, 1982].

PSL(2,7) is also the symmetry group of the Fano plane. Here it corresponds to the 168 permutations of the 7 points that carry collinear points to collinear points in the Fano plane.

Since Klein's quartic group PSL(2,7) is the automorphism group of $\mathcal{H}_{\overline{7}}^{\perp}$, which can be derived from the structure of the Fano plane, this is not too surprising. However, there is a more surprising relationship between Klein's quartic and the E_7 structures.

By way of the Thom–Arnold ADE classification of simple (complex) curve and surface singularities, we can find the correspondence (cf. Chap. 7):

$$\mathrm{E}_7 \Rightarrow X^3 + XY^3 + Z^2 \Rightarrow X^3 + XY^3$$

where X, Y, Z are basis vectors of $\mathbb{C}^3$.

The zero set of $X^3 + XY^3$ is a quartic curve with a singularity. The desingularization of this complex quartic is a genus-3 hyperbolic Riemannian surface [Barth *et al.*, 1984].

Thus there is an intimate relationship between Klein's quartic and the E_7 quartic. Both correspond in different ways to a genus-3 Riemannian surface.

Moreover, there are 28 bitangent lines attached to 56 paired points on this E_7 quartic surface [Clemens, 1980].

The symmetry group on this bitangent configuration is the E_7 Coxeter group [Hartshorne, 1977; Gray, 1982]. These 56 paired points correspond exactly to the 56 vertices of the 7D polytope corresponding to the E_7 weight lattice, and thus to the 56D fundamental representation of the E_7 Lie algebra.

There is also a close relationship between the 28 bitangents of the quartic curve, and the 27 lines of the cubic surface. This is because the automorphism group on the 27 lines is the E_6 Coxeter group, while the fundamental representation of the E_6 Lie algebra is 27-dimensional, corresponding to a 27 vertex polytope in $\mathfrak{R}^6$. Also the 54-vertex polytope in $\mathfrak{R}^6$ corresponds to the $27 + \underline{27}$ representation of E_6.

This 54 vertex polytope in $\mathfrak{R}^6$ is a substructure within the 56 vertex polytope in $\mathfrak{R}^7$, which corresponds to the 56D representation of E_7.

Moreover, the octahedral group is a subgroup (in 7 copies) of the Klein quartic group PSL(2,7) [Gray, 1982]. Thus, in summary, we have the rather intriguing relationships (cf. Chaps 2–4):

$$\begin{array}{ccccc}
\mathcal{O} & \leftarrow & \mathcal{OD} & \leftarrow & \mathbb{C}^2/\mathcal{OD} \\
\uparrow & & & & \uparrow \\
\mathrm{PSL}(2,7) & \rightarrow & \mathcal{H}_7 & \rightarrow & \mathrm{E}_7^* \text{ lattice} \\
\downarrow & & \downarrow & & \downarrow \\
\text{Fano plane} & \rightarrow & \mathcal{H}_7^{\perp} & \rightarrow & \mathrm{E}_7 \text{ lattice}
\end{array}$$

We will revisit the Fano plane structure in Chap. 14.

Chapter 14

Qubits and Black Holes

As mentioned in Chap. 13, the best quantizers up to dimension 8 are the weight lattices of the ADE Lie algebras. In this context, "quantizer" means a transformation of sets of real numbers from analog to digital form.

In general, the quantizer error rate is minimized by using higher dimensional quantizers, that is by quantizing sets of numbers rather than one number at a time. As they say, "It pays to procrastinate" [Conway and Sloane, 1988].

This can be illustrated by a list of the optimal ADE quantizers:

Dimension	Quantizer	Mean squared error
1	Z	0.08333
2	$A_2^* = A_2$	0.080188
3	$A_3 = D_3$	0.078745
	$A_3^* = D_3^*$	0.078543
4	A_4	0.078020
	A_4^*	0.077559
	$D_4^* = D_4$	0.076603
5	D_5	0.075786
	D_5^*	0.075625
6	E_6	0.074347
	E_6^*	0.074244
7	E_7	0.073231
	E_7^*	0.073116
8	$E_8^* = E_8$	0.071682

Note that, for these quantizers, the mean squared error rate diminishes as we ascend the dimensions. Also we see that in every dimension the weight lattice X_n^* performs only slightly better (if at all) than the root lattice X_n.

Thus we can consider the E_7 root lattice as the quantizing lattice, which corresponds to the dual Hamming-7 code $\mathcal{H}_7^{\perp}$, and thus to the Fano plane as described in Chap. 13.

Here we will describe the Fano plane as the structure of 7 qubits entangled as 7 sets of 3 qubits. Since this corresponds to E_7 Lie algebra structures, we will make the following conjecture:

In the case of E_7 structures, the quantizer corresponds to qubits.

Since quantizers are an aspect of classical information theory (and analog-to-digital transforms), it might seem unlikely that a particular quantizer has such an intimate relationship with the quantum information structure of qubits.

This may seem more plausible after we explore the rather special qubit structure related to E_7. The 7 qubits entangled in sets of 3 qubits are depicted in the Fano plane structure. Here 7 points are distributed into 7 lines, each containing 3 points, such that any 2 points are contained in only 1 line [Levay, 2007; Borsten *et al.*, 2008; Duff, 2010].

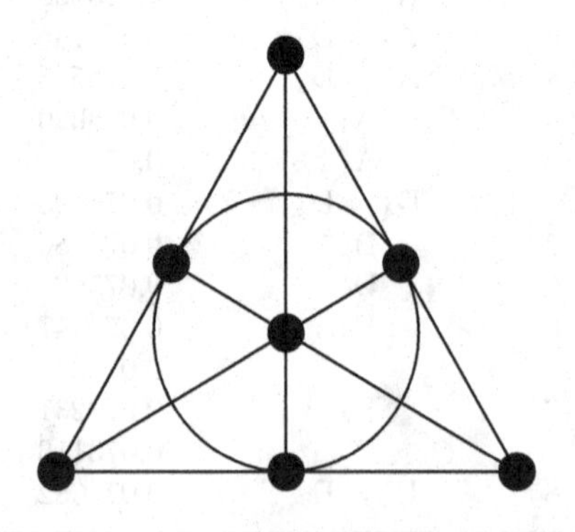

As in Chap. 13 we can label the triplets of points corresponding to the 7 lines as:

$$124, 235, 346, 457, 156, 267, 137$$

This set of 7 triplets corresponds to the dual Hamming-7 code $\mathcal{H}_7^\perp$ as shown in Chap. 13 [Sloane, 1981].

We see that the 8 codewords of $\mathcal{H}_7^\perp$ consist of the zero codeword (0000000), along with the 7 codewords corresponding to the 7 lines of the Fano plane. Here a triplet of numbers corresponds to the position of the zeros in each codeword:

124: 0010111
235: 1001011
346: 1100101
457: 1110010
156: 0111001
267: 1011100
137: 0101110

Since $\mathcal{H}_7^\perp$ can be derived from the E_7 root lattice, it is perhaps not too surprising that E_7 provides the structure for the entanglement of 7 qubits as 7 sets of 3 qubits according to the structure of the Fano plane [Duff and Ferrara, 2007].

Each qubit obeys the symmetry group $\mathcal{SL}(2, \mathbb{C})$, so that the tripartite entanglement of three quits is described by $[\mathcal{SL}(2, \mathbb{C})]^3$. Thus the entanglement of 7 qubits would be described by $[\mathcal{SL}(2, \mathbb{C})]^7$. However, in this case the 7 qubits are entangled in sets of 3 according to the structure of the Fano plane.

If we label the 7 qubits as $\{A, B, C, D, E, F, G\}$ for {Alice, Bob, Charlie, Daisy, Emma, Fred, George} as in [Duff and Ferrara, 2007], we can use the decomposition of the 56D fundamental representation of $E_{7(7)}$:

$$E_{7(7)} \supset \mathcal{SL}(2)_A \times \mathcal{SO}(6,6); 56 \rightarrow (2,12) + (1,32)$$

where $56 = 2(12) + 1(32) = 24 + 32$.

Through further decompositions we reach the decomposition:

$$E_{7(7)} \supset \mathcal{SL}(2)_A \times \mathcal{SL}(2)_B \times \mathcal{SL}(2)_C \times \mathcal{SL}(2)_D \times \mathcal{SL}(2)_E \times \mathcal{SL}(2)_F \times \mathcal{SL}(2)_G.$$

With the 56-dimensional representation decomposing as:

$$(2,2,1,2,1,1,1) + (1,2,2,1,2,1,1) + (1,1,2,2,1,2,1) \\ + (1,1,1,2,2,1,2) + (2,1,1,1,2,2,1) + (1,2,1,1,1,2,2) \\ + (2,1,2,1,1,1,2)$$

where $56 = 7(2 \times 2 \times 2) = 7 \times 8$.

Note that this decomposition perfectly matches the structure of the $\mathcal{H}_7^{\perp}$ codewords (other than the zero codeword). We merely replace 0 in each codeword with 2 in each of the $\mathcal{SL}(2)$ decompositons, while the 1's remain the same throughout this transformation.

The decomposition of $\mathrm{E}_{7(7)}$ to a maximal subgroup consisting of the product of 7 $\mathcal{SL}(2)$ groups implies the tripartite entanglement structure of 7 qubits. Each of these 7 entanglement states transforms as a (2,2,2) corresponding to 3 $\mathcal{SL}(2)$'s, and as a singlet with respect to the remaining 4 $\mathcal{SL}(2)$'s.

We can make the following convention for the Fano plane labels:

$$\begin{aligned}(2,2,1,2,1,1,1) &= 124 = |ABD\rangle \\ (1,2,2,1,2,1,1) &= 235 = |BCE\rangle \\ (1,1,2,2,1,2,1) &= 346 = |CDF\rangle \\ (1,1,1,2,2,1,2) &= 457 = |DEG\rangle \\ (2,1,1,1,2,2,1) &= 156 = |EFA\rangle \\ (1,2,1,1,1,2,2) &= 267 = |FGB\rangle \\ (2,1,2,1,1,1,2) &= 137 = |GAC\rangle\end{aligned}$$

Then we can see that these tripartite entanglement states are modeled by the multiplication table of the 7 octonions (other than the identity element).

	A	B	C	D	E	F	G
A		D	G	$-B$	F	$-E$	$-C$
B	$-D$		E	A	$-C$	G	$-F$
C	$-G$	$-E$		F	B	$-D$	A
D	B	$-A$	$-F$		G	C	$-E$
E	$-F$	C	$-B$	$-G$		A	D
F	E	$-G$	D	$-C$	$-A$		B
G	C	F	$-A$	E	$-D$	$-B$	

where, for example, the entangled state $|ABD\rangle$ corresponds to $AB = D$; $|BCE\rangle$ to $BC = E$, etc. Note also that we can read each entangled state in reverse order to produce negative octonions:

$|ABD\rangle$ corresponds to $DB = -A$;

$|BCE\rangle$ corresponds to $EC = -B$; etc. [Duff and Ferrara, 2007; Borsten *et al.*, 2008; Duff, 2010].

The octonions $\mathbb{O}$ are the elements of the largest of the normed division algebras [Dixon, 1994; Baez, 2002].

$\mathfrak{R}$ real numbers: commutative and associative (1 real dim.)

$\mathbb{C}$ complex numbers: commutative and associative (2 real dim.)

$\mathfrak{H}$ quaternions: non-commutative and associative (4 real dim.)

$\mathbb{O}$ octonions: non-commutative and non-associative (8 real dim.).

Moreover, there is a very deep relationship between these division algebras and Jordan algebras, and especially the exceptional Lie algebras. This is called the Freudenthal magic square [Baez, 2002].

	$J(\mathfrak{R})$	$J(\mathbb{C})$	$J(\mathfrak{H})$	$J(\mathbb{O})$
$\mathfrak{R}$	A_1	A_2	C_3	F_4
$\mathbb{C}$	A_2	$A_2 + A_2$	A_5	E_6
$\mathfrak{H}$	C_3	A_5	D_6	E_7
$\mathbb{O}$	F_4	E_6	E_7	E_8

Note that each Jordan algebra $J(k)$ is the algebra of 3×3 hermitian matrices over k.

Note also that the exceptional Lie group G_2 is the automorphism group of $\mathbb{O}$, so that all five exceptional Lie algebras $(G_2, F_4, E_6, E_7, E_8)$ are octonian symmetry structures.

The split forms of these Lie algebras are implicated in a magic square of the split composition algebras. A split Lie algebra has non-compact dimensions as well as compact dimensions.

For example $\mathfrak{e}_{7,(7)}$ is a split form of E_7 in which the difference between the number of non-compact dimensions and the number of compact dimensions is 7. In this case, the 133 dimensions of E_7 become 63 compact dimensions plus 70 non-compact dimensions. The 63 compact dimensions belong to the maximal compact subalgebra $\mathfrak{su}_8$. Note that $70 - 63 = 7$.

The split version of the magic square becomes:

	$J(\mathfrak{R})$	$J(\mathbb{C}^s)$	$J(\mathfrak{H}^s)$	$J(\mathbb{O}^s)$
$\mathfrak{R}$	$\mathfrak{so}_3$	$\mathfrak{sl}_3(\mathfrak{R})$	$\mathfrak{sp}_6(\mathfrak{R})$	$\mathfrak{f}_{4(4)}$
$\mathbb{C}^s$	$\mathfrak{sl}_3(\mathfrak{R})$	$\mathfrak{sl}_3(\mathfrak{R}) \oplus \mathfrak{sl}_3(\mathfrak{R})$	$\mathfrak{sl}_6(\mathfrak{R})$	$\mathfrak{e}_{6(6)}$
$\mathfrak{H}^s$	$\mathfrak{sp}_6(\mathfrak{R})$	$\mathfrak{sl}_6(\mathfrak{R})$	$\mathfrak{so}_{6,6}$	$\mathfrak{e}_{7(7)}$
$\mathbb{O}^s$	$\mathfrak{f}_{4(4)}$	$\mathfrak{e}_{6(6)}$	$\mathfrak{e}_{7(7)}$	$\mathfrak{e}_{8(8)}$

Note that the quadratic forms of the split composition algebras have split signatures: $\mathbb{C}^s$(1,1); $\mathfrak{H}^s$(2,2); $\mathcal{O}^s$ (4,4). For example, the $\mathbb{C}^s$ quadratic form is $x^2 - y^2$, which implies that for $x + yj, j^2 = +1$, replacing $i^2 = -1$ in $\mathbb{C}$.

Note also that the non-compact group $\mathrm{E}_{7(7)}$ has long been viewed as a symmetry group for 11D supergravity theory [Hull, 1985; de Wit and Nicolai, 1986]. This theory is called $N = 8$ supergravity and has the following particle content:

1 graviton (spin-2) — $\mathcal{SU}(8)$ singlet
8 gravitini (spin-3/2) — fundamental $\mathcal{SU}(8)$ representation
28 vector fields (spin-1) — adjoint representation of $\mathcal{SO}(8) \subset \mathcal{SU}(8)$
56 fermions (spin-$\frac{1}{2}$) — fundamental $\mathrm{E}_{7(7)}$ representation
70 scalars (spin-0) — coset space $\mathrm{E}_{7(7)}/\mathcal{SU}(8)$

This theory is called $N = 8, D = 4$ supergravity because the 11-spacetime dimensions are compactified on the 7D torus, which is the Cartan subgroup of $\mathrm{E}_{7(7)}$. The uncompactified dimensions constitute 4D Minkowski space.

Note, however, that 11D supergravity is the "low energy" approximation to type **II** superstring theories. This is because the type **II** string theory compactified on T^6 has an effective action which is equivalent to that of $N = 8$ supergravity compactified on T^7 [Hull and Townsend, 1995].

Alternatively, the 11D spacetime can be compactified on S^7, with the uncompactified dimensions being AdS_4 (cf. Chap. 15).

One notes immediately that S^7 can be viewed as the unit length octonions, which accommodates the octonionic structures described previously in this chapter. Here we could contemplate $AdS_4 \times S^7$ as the 11D spacetime of M-theory compactified on S^7. The AdS/CFT duality implicit in this structure describes gravity in the noncompact spacetime AdS_4, while the dual theory is the superconformal field theory on the 3D boundary at infinity of AdS_4 [Aharony, Bergman, Jeffries and Maldacena, 2008].

However, we will defer a deeper description of the AdS/CFT duality to Chap. 15.

Here we will describe another rather striking duality: qubit triplets correspond to extreme black holes [Duff, 2007, 2010; Kallosh and Linde, 2006].

Extreme black holes are unlike the classical black holes in several ways. The classical black hole has all its mass confined within a radius R such that even light cannot escape. This radius, as calculated by Schwarzschild [1916] is, cf. [Misner, Thorne, and Wheeler, 1973]:

$$R_s = 2GM/c^2,$$

where G is Newton's gravity constant, M is the confined mass, and c is the speed of light.

Many black holes (of 20 solar masses or more) have been discovered. Moreover, there is strong evidence for an extremely massive black hole at the center of every galaxy including the Milky Way.

These are called classical black holes because they are based on the Schwarzschild solution to Einstein's field equations for general relativity. However, the work of [Hawking, 1974] has shown that quantum theory modifies the picture of classical black holes so that they radiate via Hawking radiation, and thus have a temperature and an entropy related to the area of the event horizon of the black hole. The rather amazing formula for this entropy relationship is:

$$S/k = 2\pi Ac^3/4\hbar G = A/4\mathcal{L}^2$$

where S is entropy, k is Boltzmann's constant, A is the area of the event horizon, c is the speed of light, $\hbar$ is Planck's constant, G is Newton's constant, and $\mathcal{L}^2$ is the Planck area.

Note that $\mathcal{L} =$ is the Planck length $(\hbar G/2\pi c^2)^{1/2} = 1.6\times10^{-33}$ cm so that the entropy of a stellar (or galactic) black hole would be extremely large.

The black hole temperature has the formula:

$$kT = \hbar c^3/16\pi^2 GM,$$

where M is the black hole mass. Thus the smaller the black hole, the greater would be its temperature and the energy of its Hawking radiation [Zwiebach, 2004].

Since the ordinary matter of a stellar or galactic black hole would have equal positive and negative electric charges, such classical black holes would carry no net electric charge.

However, the Reissner–Nordström solution describes a charged black hole. In this case we can consider extreme black holes which are defined as those whose electric charge Q is as large as can be accommodated by its mass M as confined within its event horizon [Becker *et al.*, 2007].

Such black holes cannot have a naked singularity, which would be the case if its mass were less than its charge. Of special interest are extreme black holes which saturate its BPS bound (named for Bogomolny, Prasad and Summerfield) defined by:

BPS bound: $M \geq |Q|$,

BPS black hole: $M = |Q|$,

Non-BPS black hole: $M > |Q|$.

The BPS black holes in the context of quantum gravity are supersymmetric and can be described in the context of supergravity and string theory.

For example, in the heterotic string theory its 10D fermionic dimensions can be compactified on T^6, so that 4D Minkowski space remains uncompactified. Then T^6 accommodates three complex scalar fields, conventionally called S, T and U. Each of these scalar fields has an $\mathcal{SL}(2, \mathbb{Z})$ duality group along with a triality permuting S, T and U.

Moreover, these symmetries constrain the entropy of extremal black holes, with 8 charges (4 electric, 4 magnetic) which can be described by a $2 \times 2 \times 2$ hypermatrix so that the squared entropy corresponds to Cayley's hyperdeterminant [Duff, 2007, 2010].

As a rather amazing consequence, there is a 1–1 correspondence between sets of 3 qubits of quantum information theory and the 3 fields (S, T and U) of extremal black holes.

In fact, the 8 black hole charges can be identified with 8 numbers that fix the 3-qubit state. Thus, if we call the 3 qubits A, B, C, (for Alice, Bob and Charlie) we can identify these qubits with S, T and U. This is because the entropy of these extreme black holes is given by the 3-tangle of the qubits which also corresponds to Cayley's hyperdeterminant $\mathfrak{h}$. This hyperdeterminant corresponds to three classes of 3-tangles, and thus to the black hole squared entropies:

$$S^2 = \mathrm{Det}\mathfrak{h}$$

Three classes: (1) Det $\mathfrak{h} < 0$, (2) Det $\mathfrak{h} = 0$, (3) Det $\mathfrak{h} > 0$

(1) Non-separable tripartite (GHZ) class $\Leftrightarrow N = 4$ black hole (1/8 SUSY)
(2) Tripartite (W) class $\Leftrightarrow N = 3$ black hole (1/8 SUSY)
Bipartite A-BC class $\Leftrightarrow N = 2$ black hole (1/4 SUSY)
Bipartite B-CA class $\Leftrightarrow N = 2$ black hole (1/4 SUSY)
Bipartite C-AB class $\Leftrightarrow N = 2$ black hole (1/4 SUSY)
Separable A-B-C class $\Leftrightarrow N = 1$ black hole (1/2 SUSY)
(3) Non-separable tripartite GHZ(2) class $\Leftrightarrow N = 4$ non-BPS black hole (an extreme black hole that exceeds the BPS bound).
Note that the BPS bound is saturated whenever one half, one quarter, or one eighth of the supersymmetry (SUSY) generators remain unbroken. For 4 uncompactified spacetime dimensions, the spinor representation has 4 real components, so the number of supersymmetries (before breaking) is 4N [Polchinski 1998, Vol. II].

$N = 4$: 16 original SUSYs are broken to 2: (1/8 SUSY)

$N = 2$: 8 original SUSYs are broken to 2: (1/4 SUSY)

$N = 1$: 4 original SUSYs are broken to 2: (1/2 SUSY)

The relationship between the local qubit entropies $\mathcal{S}_i$ and the black hole entropies, $\mathcal{S}^2 = \text{Det}\ \mathfrak{h}$, can be described by the following table [Borsten *et al.*, 2008; Duff, 2010]:

Type	$\mathcal{S}_A$	$\mathcal{S}_B$	$\mathcal{S}_C$	Det $\mathfrak{h}$	Black hole	SUSY	N
GHZ	>0	>0	>0	>0	large	1/8	4
W	>0	>0	>0	0	small	1/8	3
A-BC	0	>0	>0	0	small	1/4	2
B-CA	>0	0	>0	0	small	1/4	2
C-AB	>0	>0	0	0	small	1/4	2
A-B-C	0	0	0	0	small	1/2	1
GHZ(2)	>0	>0	>0	>0	large	0	4

Both GHZ and W states are full 3-qubit entanglement states. The GHZ state is fragile in the sense that a loss of one qubit causes the GHZ state to descend to the unentangled A-B-C state. However, W is more robust, because a loss of one qubit causes the W state to descend to one of the 2-qubit ABC states [Dur, Vidal, and Cirac, 2000; Kallosh and Linde, 2006].

Here GHZ(2) corresponds to a black hole that is non-BPS since its mass exceeds its charge: $M > |Q|$, and Det $\mathfrak{h} > 0$.

Note that "large" black hole in this classification means that the classical description of the black hole horizon is finite (i.e. not equal to zero). The "small" black hole is one whose classical horizon is zero; yet it acquires a non-zero size due to quantum corrections via supergravity.

The $N = 8$ case is very special, because (as described above), this corresponds to 7 qubits organized as 7 sets of 3 qubits corresponding to the Fano plane. The corresponding black holes are classified by the sign of the quartic $E_{7(7)}$ Cartan invariant I_4, which magically happens to be equal to the negative of the Cayley's hyperdeterminant [Duff and Ferrara, 2007]:

$$I_4 = -\text{Det}\ \mathfrak{h}.$$

So these $N = 8$ black holes are also classified by the sign of Det $\mathfrak{h}$.

(1) $I_4 > 0$ (Det $\mathfrak{h} < 0$) $\Leftrightarrow$ large black holes (BPS): 1/8 SUSY
(2) $I_4 = 0$ (Det $\mathfrak{h} = 0$) $\Leftrightarrow$ small black holes (BPS): 1/8, 1/4, 1/2 SUSY
(3) $I_4 < 0$ (Det $\mathfrak{h} > 0$) $\Leftrightarrow$ large black holes (non-BPS) non-SUSY.

Recent investigations have been able to describe the duality between black holes and 4-qubit entanglements. [Levay, 2007] uses a supergravity description to extend the STU 3-qubit classification to 4-qubits. [Borsten *et al.*, 2008] use string theory to provide a classification of 9 families of 4-qubit/black hole duality. However, this will not be described here.

In this chapter we have emphasized the role of E_7 especially in its $E_{7(7)}$ form. Thus it is useful to put $E_{7(7)}$ in the larger context of the ADE type symmetry groups in (for example) Type **II** string theories. Thus we can display the following table [Hull and Townsend, 1995]:

Dim.	Supergravity group G	T-duality	U-duality	ADE type
$10A$	$\mathcal{SO}(1,1)/\mathbb{Z}_2$	$\mathbf{1}$	$\mathbf{1}$	A_0
$10B$	$\mathcal{SL}(2,\mathfrak{R})$	$\mathbf{1}$	$\mathcal{SL}(2,\mathbb{Z})$	A_1
9	$\mathcal{SL}(2,\mathfrak{R}) \otimes \mathcal{O}(1,1)$	$\mathbb{Z}_2$	$\mathcal{SL}(2,\mathbb{Z}) \otimes \mathbb{Z}_2$	A_1
8	$\mathcal{SL}(3,\mathfrak{R}) \otimes \mathcal{SL}(2,\mathfrak{R})$	$\mathcal{O}(2,2,\mathbb{Z})$	$\mathcal{SL}(3,\mathbb{Z}) \otimes \mathcal{SL}(2,\mathbb{Z})$	$A_2 \oplus A_1$
7	$\mathcal{SL}(5,\mathfrak{R})$	$\mathcal{O}(3,3,\mathbb{Z})$	$\mathcal{SL}(5,\mathcal{Z})$	A_4
6	$\mathcal{O}(5,5)$	$\mathcal{O}(4,4,\mathbb{Z})$	$\mathcal{O}(5,5,\mathbb{Z})$	D_5
5	$E_{6(6)}$	$\mathcal{O}(5,5,\mathbb{Z})$	$E_{6(6)}(\mathbb{Z})$	E_6
4	$E_{7(7)}$	$\mathcal{O}(6,6,\mathbb{Z})$	$E_{7(7)}(\mathbb{Z})$	E_7
3	$E_{8(8)}$	$\mathcal{O}(7,7,\mathbb{Z})$	$E_{8(8)}(\mathbb{Z})$	E_8
2	$E_{9(9)}$	$\mathcal{O}(8,8,\mathbb{Z})$	$E_{9(9)}(\mathbb{Z})$	E_9
1	$E_{10(10)}$	$\mathcal{O}(9,9,\mathbb{Z})$	$E_{10(10)}(\mathbb{Z})$	E_{10}

For the heterotic string theories, the supergravity group G, is even more closely related to the T-duality and U-duality groups. These groups are all of D-type, running from $D8$ through $D16$ [Kaku, 1999].

Note: *T-duality* in a toroidally compactified string theory is an invariance between the winding number of a string and the momentum number of the string. Thus the size of the tori does not affect the predictions of the theory.

Also the mirror symmetry between pairs of Calabi–Yau spaces is equivalent to T-duality [Becker *et al.*, 2007] cf. Chap. 16.

U-duality is a symmetry between large toroidal radii with strong coupling and small toroidal radii with weak coupling.

In general, symmetries are described by symmetry groups, so that here we have a series of (dimension dependent) symmetry groups of ADE type.

Chapter 15

The Holographic Principle

In 1974 three papers in different areas of physics were published. However, they became three intertwined paths to the statement and development of the Holographic Principle in the '90s.

(1) Stephen Hawking claimed that black holes evaporate: "Black hole explosions?" [Hawking, 1974]. As he put it in *A Brief History of Time*, "Black holes ain't so black" [Hawking, 1988].

Two years later, Hawking announced the "paradox" that, because black holes evaporate, they destroy information and thus contradict quantum mechanics [Hawking, 1976]. This led to a crisis in theoretical physics, which was resolved 20 years later by way of the holographic principle [Susskind, 1995, 2003].

(2) Gerard 't Hooft proposed that the $\mathcal{SU}(3)$ gauge group for quantum chromodynamics (QCD) be replaced by a series of $\mathcal{SU}(N)$ groups where N is the number of colors and approaches infinity: $N \to \infty$. He claimed that this would facilitate calculations in the theory of strongly interacting particles, as spinning strings ['t Hooft, 1974].

It should be remembered that string theory began as an attempt to describe the strongly interacting particles called hadrons. For example, a meson could be described as a string spinning at light speed and with a quark and antiquark at each end of the string, of length about a Fermi = 10^{-13} cm [Susskind, 1970; Nambu, 1970].

(3) Joel Scherk and John Schwarz, took seriously the fact that the spinning string contained, as one of its vibrational states, the massless spin-2 particle. No such strongly interacting particle was known, but it had long been known that any attempt to

quantize gravity would entail the graviton as a massless spin-2 particle. Accordingly, Scherk and Schwarz proposed that string theory be considered primarily as a theory of quantum gravity, with strings not much larger than the Planck length, 10^{-33} cm [Scherk and Schwarz, 1974; Schwarz, 1985].

The quantum-gravity string theory of Scherk and Schwarz was for ten years eclipsed by the discovery in 1976 of supergravity [Freedman *et al.*, 1976; van Nieuwenhuizen and Freedman, 1979]. This was a point particle theory, whose most supersymmetric version had 11 spacetime dimensions and $N = 8$ supersymmetries. The 11 dimensions could be compactified on the 7D torus, leaving 4 uncompactified dimensions. Thus it would have global E_7 invariance and $\mathcal{SU}(8)$ local invariance [Howe, 1979; van Nieuwenhuizen, 1980].

However, in 1984, the string theory version of quantum gravity (with hope of including all the particles of the standard model) came back into focus [Green and Schwarz, 1984; Gross *et al.*, 1985].

By 1985 there were five competing superstring theories, each a self-consistent model of quantum gravity. Three theories (called I, IIA and IIB) were 10D spacetime theories. Two theories were called heterotic string theories, because they were a subtle interweaving of the original 26D bosonic theory and the 10D fermionic theory. The symmetry groups for these heterotic theories were $\mathcal{SO}(32)$ (i.e. D_{16}) and $E_8 \otimes E_8$.

Moreover, E_8 and D_{16} can be projected to lower dimensional ADE groups:

$$E_8 \to E_6 \to D_5 \to A_4; \quad D_{16} \to D_5 \to A_4,$$

where D_5 is the Coxeter label corresponding to $\mathcal{SO}(10)$, and A_4 is the Coxeter label for $\mathcal{SU}(5)$. Both $\mathcal{SU}(5)$ and $\mathcal{SO}(10)$ were proposed as grand unified theories (GUTs) by [Georgi and Glashow, 1974] so that GUTs were another idea that went back to that magic year 1974 [Georgi, 1982; Zee, 1982].

Note that in the GUT projections E_7 was skipped over for certain technical reasons, but came back into the picture via 11D supergravity as mentioned above.

The problem of the five competing string theories was brilliantly solved by Edward Witten in 1995, when he showed that an overarching theory, which he called M-theory unified the five superstring theories via certain dualities (called S and T). Since M-theory is an 11D spacetime theory, it has 11D supergravity as a "low energy" limiting theory [Witten, 1995].

These duality mappings were depicted in a diagram in Chap. 8. For clarity, we copy it below from Chap. 8 (for further details).

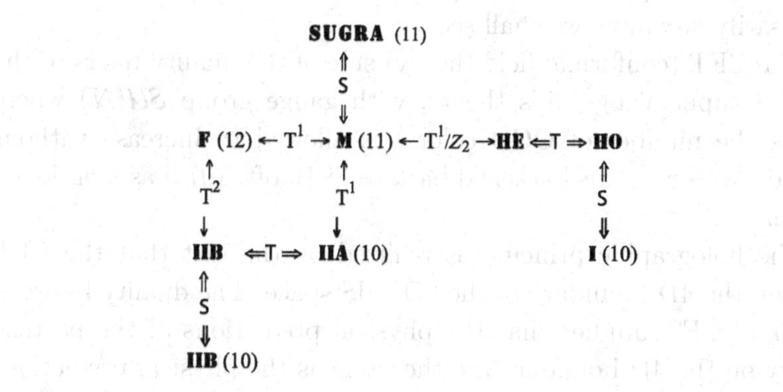

The holographic principle was conjectured by 't Hooft in 1993 and further developed by Susskind (who had been visiting 't Hooft). This was stated in terms of information storage ['t Hooft, 1993; Susskind, 1995].

The maximum amount of information in any volume of space is less than the amount of information that can store on the boundary area of that volume of space (using one bit per 4 Planck areas) [Susskind, 2005].

This was first described in the context of the 2D horizon area of a black hole with respect to the 3D interior of the black hole, in order to deal with Hawking's claim that any information falling into a black hole would be lost forever. Since this would contradict the unitarity principle of quantum mechanics, Hawking viewed this loss of information as a crisis in physics.

Susskind's paper on the holographic principle was published in 1995, the same year Witten unified the five competing 10D superstring theories by way of dualities and embeddings in 11D M-theory, a revolutionary achievement.

Only two years later, Juan Maldacena stunned the physics world again by proposing the AdS/CFT duality, which was a concrete realization of the holographic principle, as pointed out in detail in [Maldacena, 1998; Witten, 1998].

The AdS side of the duality refers to the 10D spacetime made up of 5D anti-de Sitter space and the 5D sphere (as the compactified part) of IIB superstring theory. Note that S^5 is compact but not necessarily small, as we shall see.

The CFT (conformal field theory) side of the duality refers to the $\mathcal{N} = 4$ super Yang–Mills theory, with gauge group $\mathcal{SU}(N)$ where N (as the number of QCD colors) is allowed to increase without bound: $N \to \infty$. This harkened back to ['t Hooft, 1974] as mentioned above.

The holographic principle is realized by the fact that the CFT lives on the 4D boundary of the 5D AdS space. The duality between AdS and CFT implies that the physical predictions of the particle theory on the 4D boundary are the same as the physical predictions of the IIB superstring quantum gravity theory on the $\mathrm{AdS}_5 \otimes \mathrm{S}^5$.

The big caveat here is that, although this is a mathematically robust result, the AdS space where gravity reigns (in quantum gravity form) is not only a 5D spacetime but is negatively curved, whereas cosmological observations show that the universe is very slightly positively curved. This is ordinarily accounted for by an extremely small cosmological constant in Einstein's [1917] cosmological field equation [Misner *et al.*, 1973]. Moreover, this cosmological constant is the preferred description of the so-called dark energy causing the slight increase in the acceleration of the growth of the universe (rather than a decrease, as was widely expected) [Weinberg, 2008].

There are, of course, attempts to provide a de Sitter space version of the holographic principle, dS/CFT, which would fit the established cosmology.

Here, however, we will concentrate on the mathematically robust holographic duality already established by Maldacena, Witten and many other physicists.

In keeping with our emphasis on the ADE structures, we can immediately point out that the infinite series of $\mathcal{SU}(N)$ gauge groups describing the CFT living on the 4D boundary of AdS_5 can be viewed as a series of A type groups. Note that $\mathcal{SU}(N)$ is a Lie group of type A_{N-1}.

Also note that the S^3 boundary of the spatial part of 5D AdS is the geometrical structure of the $\mathcal{SU}(2)$ Lie group (which can be seen as the set of unit length quaternions). Moreover, as pointed out in Chap. 4, the set of finite subgroups of $\mathcal{SU}(2)$, which I call McKay groups, are ADE classified. Accordingly, I conjecture that these McKay groups will have a role to play in the AdS/CFT duality.

In order to better describe this S^3 boundary, we can depict a 3D sub-geometry of the 5D AdS. This 3D AdS can be viewed as a cylinder whose 2D disk cross-section is considered as a conformal mapping of a hyperbolic plane. Thus the S^1 boundary of the disk is the infinitely distant boundary of the hyperbolic plane. The length of the cylinder is the time parameter t, and is itself infinitely long.

Similarly in the 5D AdS case, the bulk space is the 4D ball, B^4 (analogous to the disk) attached to each point of time t. Thus the boundary at infinity of B^4 is the sphere S^3. It must be understood that B^4 is a conformal mapping of a 4D hyperbolic space (with negative curvature).

The metric for 5D AdS via [Susskind and Lindesay, 2005] is:

$$d\tau^2 = R^2/(1-r^2)^2\{(1+r^2)^2dt^2 - 4dr^2 - 4r^2d\Omega^2\},$$

where t is time, r is the radial coordinate $(0 \leq r < 1)$, and Ω provides parameters for S^3. R is the radius of curvature so that the constant negative curvature is R^{-2}.

There are alternative metrics, but the above metric describes a geodesically complete geometry, rather than a local patch of AdS_5, In this alternative case S^3 is replaced by a 3D tangent plane $\mathfrak{R}^3$, so that the inclusion of t results in the geometry of 4D Minkowski space.

In order to describe the full 10D space, $AdS_5 \otimes S^5$, we must add to the above metric the 5-sphere metric:

$$R^2d(\omega_5)^2.$$

It is amusing to note that Einstein's original (1917) cosmology was described as the cylindrical spacetime, $t \otimes S^3$ [Eddington, 1923].

Einstein added the cosmological tensor with constant λ to his field equations in order to keep S^3 static throughout time, because in 1917 the universe (according to astronomers) was nothing more than the Milky Way galaxy. Einstein's cosmological constant λ had to be a small positive number in order to act as a repulsive force counteracting the gravitational attractive force between all the masses in the universe.

In the '20s Edwin Hubble showed that the universe has many galaxies beyond the Milky Way [Misner *et al.*, 1973]. Moreover, the increasing redshifts of distant galaxies showed that the universe is expanding. As a consequence, Einstein dropped his cosmological tensor from his cosmological equation.

In an ironic reversal, recent measurements of very distant redshifts show that the expansion of the universe is accelerating. This can be modeled by the inclusion of a very tiny positive cosmological constant [Weinberg, 2008].

An alternative cosmology, due to Willem de Sitter (1917), assumes that the universe contains no matter at all [Weinberg, 2008]. This could have been an appropriate approximation because the average matter density of the universe is very close to zero. The de Sitter universe is spherical, S^4, with the understanding that time is *imaginary*, so that the Minkowski spacetime metric is replaced by the de Sitter metric (where $\tau = it$):

$$x^2 + y^2 + z^2 - t^2 \Rightarrow x^2 + y^2 + z^2 + \tau^2 .$$

The de Sitter metric is equivalent to the Wick rotated metric used in quantum field theory, in which the Wick rotation in the complex plane transforms the real number line into the imaginary number line.

However, in 4D de Sitter space, the *real* time coordinate describes hyperbolas rather than circles [Eddington, 1923; Hawking, 2001].

Recently Hawking has invoked S^4 with imaginary time as the cosmic geometry in order to obviate the problem of the singularity at the beginning of time [Hartle and Hawking, 1983; Gibbons, 2003; Page, 2003].

As a consequence of Hubble's observation of the expansion of the universe, de Sitter (like Einstein) abandoned the cosmological constant λ.

Now that we have observational evidence for a very tiny positive λ, there is renewed interest in de Sitter space, dS_4, which is thought to approximate both early and late phases of the universe.

However, in this chapter we are concerned with the anti-de Sitter space, AdS_5. Moreover, it is indeed curious that the 4D boundary at infinity of AdS_5 has the geometrical structure of Einstein's cylindrical universe:

$$t \otimes S^3.$$

Note that this geometry is indeed static, since it is the boundary at infinity of AdS_5. It is this 4D boundary that is the arena for the CFT in the AdS/CFT duality. Moreover, the holographic principle claims that the information stored in the boundary, $t \otimes S^3$, is equivalent to the information in the entire AdS_5 bulk geometry.

Another model for the full 10D spacetime, $AdS_5 \otimes S^5$, is an analog to an extreme Reissner–Nordström (R–N) black hole. Such a black hole has its electric charge equal to its mass, so that it has only one event horizon. This event horizon can be depicted as a circle somewhere along an infinitely long trumpet shaped throat. This circle is a 1D submanifold of the black hole horizon S^2. The trumpet shape is negatively curved, so that it represents a 2D AdS space. Thus the full geometry of the extreme R–N black hole is according to [Warner, 2003]:

$$AdS_2 \otimes S^2 \text{ a submanifold of } AdS_5 \otimes S^5.$$

As a trumpet shaped submanifold, the R–N black hole provides an analog of the 10D spacetime. Thus AdS_5 has a 5D trumpet shaped (negatively curved) geometry, with an infinity of S^5 spheres orthogonal to the 5D AdS trumpet-like geometry. These S^5 spheres are analogous to the infinity of circles orthogonal to the 2D trumpet geometry AdS_2.

The boundary at infinity of AdS_5 is made up of a stack of 3D Dirichlet branes (called D3-branes) where D3 is to be distinguished from the Coxeter graph label D_3). These D3-branes are 3D membranes which can have open strings attached to them. These strings correspond to particles other than the graviton (which must correspond to a closed string), so they do not feel the gravitational force. They are free to move and interact on the D3-branes and also between the D3-branes (but not in the 5D or 10D bulk). The gravitons, as closed strings, are free to move in the 10D bulk space $AdS_5 \otimes S^5$.

In this picture the number N of D3-branes is equivalent to the order of the QCD-like $\mathcal{SU}(N)$ gauge group. So this is where the series of $\mathcal{SU}(N)$s with N increasing comes into play. These $\mathcal{SU}(N)$s correspond to the infinite series of A_{N-1} Coxeter graphs.

The stack of D3-branes induce an extra dimension which can be interpreted as time. Thus we have here the structure of the 4D boundary of the AdS_5 geometry.

The claim of the AdS/CFT duality in this case is that the 4D boundary made up of the D3-branes (and their attached strings) contains the same information as the 5D geometry AdS_5. This, of course, carries with it many implications, which provide a kind of dictionary for the AdS/CFT duality.

(1) The AdS/CFT is the most mathematically precise version of the holographic principle. This is because of the duality of information between the boundary and the bulk space.

(2) This AdS/CFT duality is an S-duality, which means that when the CFT has a strong coupling, the AdS has weak coupling; and vice versa.

Thus this duality makes it possible to perform calculations in the weak coupling AdS side equivalent to the intractable CFT calculations with strong coupling for large N.

(3) The CFT Yang–Mills coupling constant g_{YM} corresponds to the string coupling constant g_s, so that $g_{\mathrm{YM}}^2 = 4\pi g_s$. This accords with the fact that each of the AdS/CFT dualities has two dimensionless parameters.

For AdS: $R/\mathcal{L}$ is the radius of curvature measured in string lengths $\mathcal{L}$.

g_s is the string coupling constant.

For CFT: N is order of the Y–M gauge group $\mathcal{SU}(N)$.

g_{YM} is the Y–M gauge coupling constant.

These relationships imply that the bulk parameters R and g_s can be derived from the CFT boundary parameters:

$$R/\mathcal{L} = (N g_{YM}^2)^{1/4},$$
$$g_{YM}^2 = 4\pi g_s \text{ (as mentioned above)}$$

[Susskind and Lindesay, 2005].

(4) The CFT is a supersymmetric Yang–Mills $\mathcal{SU}(N)$ theory with $N = 4$ supersymmetry. The AdS is also supersymmetric with $N = 8$, which is the maximum supersymmetry for supergravity and superstring theory.

(5) The isometry group of the superstring theory in the 10D bulk space is the same as the superconformal symmetry group of the $N = 4$ Yang–Mills gauge theory on the 4D boundary of AdS_5. This group is called PSU(2, 2|4), where $\mathcal{SU}(2,2)$ is the conformal group, and PSU(2, 2|4) means that there are 4 supersymmetry generators, while the subgroup is $\mathcal{SU}(2,2) \times \mathcal{SU}(4)$. Note that AdS_5 has isometry group $\mathcal{SO}(4,2)$ whose covering group is $\mathcal{SU}(2,2)$, while S^5 has isometry group $\mathcal{SO}(6)$ whose covering group is $\mathcal{SU}(4)$. The covering groups are necessary because supersymmetry entails fermions. The P in PSU(2, 2|4) means that an extra $\mathcal{U}(1)$ group has been projected out of the subgroup, because $2 + 2 = 4$ [Becker *et al.*, 2007].

On the CFT side of the duality, $\mathcal{SU}(4)$ is the global symmetry of the $N = 4$ Yang–Mills theory. The supersymmetries do not commute with $\mathcal{SU}(4)$, but they do form commutators with the conformal symmetry $\mathcal{SU}(2,2)$, and thus account for the extra algebraic structure which leads to the superconformal algebra of PSU(2, 2|4).

(6) The N coincident D3-branes at the boundary of AdS_5 carry N units of charge. This charge has a five-form flux which is measured by enclosing the D3-branes in the S^5 sphere in the 10D space $AdS_5 \otimes S^5$. Thus $\mathcal{SU}(N)$ plays a role dual to the isometry group $\mathcal{SU}(4)$ for S^5 in the 10D spacetime of the IIB string theory.

(7) For the AdS/CFT duality to hold, N in $\mathcal{SU}(N)$ must become extremely large. As a consequence, the stack of D3-branes achieves a mass equal to its charge. This extreme N stack of D3-branes becomes

an extreme (R–N) black hole. These branes are then called black-branes, cf. Chap. 14.

The near horizon of this black-brane stack is warped by the high mass-density of this R-N extreme black hole. The geometry of this warped bulk space is $AdS_5 \otimes S^5$. The radius of curvature R (in Planck units) is equivalent to a positive power of 1/N [Maldacena, 1998; Becker *et al.*, 2007].

The holographic principle is unique in the sense that it is the basis of the only experimental test of string theory. This test entailed a string theory graviton calculation in AdS_5 in order to determine the viscosity of a "quark-gluon plasma" living on the 4D boundary of AdS_5.

This quark-gluon plasma was created at the Relativistic Heavy Ion Collider (RHIC) at the Brookhaven National Laboratory. In this experiment, heavy ions such as gold nuclei are smashed into each other at high velocities so that the quarks and gluons making up the protons and neutrons interact with each other at very high energies. It was assumed by the particle physicists that a plasma (an ionized gas) of quarks and gluons would be created. The gauge theory of quantum chromodynamics (QCD) implies that quarks interacting closer to each other experience a weaker QCD force. This is called asymptotic freedom and accounts for the trapping of quarks inside protons and neutrons. Thus when the quarks are farther apart they experience a stronger QCD force. This was discovered in 1973 and led to the widespread acceptance of the quark model of the strong nuclear force modeled by the gauge group $\mathcal{SU}(3)$, where three is the number of quark charges called "color" [Gross and Wilczek, 1973; Politzer, 1973; 1974].

Quite amazingly, however, rather than a gas-like state the RHIC experiment produced a state in which the quarks are strongly coupled and behave like a liquid with extremely low viscosity — a kind of perfect fluid.

This, of course, made no sense in terms of the Standard Model QCD [Blau, 2005; Son, 2005].

Even more amazingly, string theory could model this perfect fluid behavior by describing the properties of a black hole embedded in a

5D spacetime, rather than an ordinary black hole embedded in the usual 4D spacetime.

This string theory calculation is based on a holographic duality between a gravity theory acting in the bulk space AdS_5 and a kind of QCD theory acting on the 4D boundary. This is not the AdS/CFT duality as described earlier in this chapter. Rather this duality could be described as:

$$\text{AdS/QCD where } N \to \infty.$$

Thus on the boundary there exists a theory which is neither conformal (such as CFT) nor supersymmetric. However the N in $\mathcal{SU}(N)$ must be large (approaching infinity) as in the ['t Hooft, 1974] paper.

Such a theory would describe an extremely strong coupling and thus be impossible to calculate using the usual perturbation methods of Feynman diagrams. Even ordinary QCD with $\mathcal{SU}(3)$ is extremely difficult to calculate, so that with higher $\mathcal{SU}(N)$s the calculations become completely impossible.

However, with the weak/strong S-duality of the holographic principle, we can expect that strong couplings on the boundary geometry corresponds to weak coupling in the bulk geometry.

In the case of the RHIC experiment, the bulk calculations were made by [Policastro *et al.*, 2001]. They showed that the boundary behavior could be modeled by gravitons traveling from a point on the boundary to bounce off (i.e. scatter from) a black hole in the bulk and travel back to another point on the boundary. These gravitons are modeled as closed strings, so that string theory is fundamentally entailed. Also due to the strangely warped behavior of AdS_5, with its negative curvature this trip from boundary to black hole to boundary occurs in finite time.

In their calculation the scattering cross section is proportional to the area of the black hole horizon and also to the viscosity of the interactions on the boundary. This viscosity would be characterized by:

$$\eta/s,$$

where η is the shear viscosity, and s is the entropy density.

Then, since the entropy of a black hole is proportional to the area of the horizon, the two black hole areas mentioned cancel against each other cf. [Hawking, 1974].

Thus one can derive:

$$\eta/s = (1/4\pi)\hbar/k,$$

where $\hbar$ is Planck's constant divided by 2π, and k is Boltzmann's constant cf. [Johnson and Steinberg, 2010].

In [Kovtun *et al.*, 2005] a universal lower bound for this viscosity measure, η/s was proposed. This was plausible because water and even the superfluid liquid helium have values far above η/s, whereas the RHIC quark/gluon state has a viscosity only slightly higher than η/s.

More recent theoretical work has revealed systems with viscosity measures even smaller than η/s. However, such systems have not been seen in laboratory experiments. It may be that η/s should be described as a universal element of an expansion in $1/N$, as in the ['t Hooft, 1974] paper, where N is the number of colors in the infinite series of $\mathcal{SU}(N)$s, which is to say the infinite series of A_{N-1}.

Chapter 16

Calabi–Yau Spaces and Mirror Symmetry

In 1954 Eugenio Calabi put forward the surprising conjecture that there exist complex n-dimensional, compact manifolds with zero Ricci curvature [Calabi, 1954; 1955].

This conjecture was generally assumed to be false because it seems to require a specific geometric condition to follow from a very general topological condition.

Shing-Tung Yau also doubted Calabi's conjecture and in fact thought he could disprove it by finding a counterexample. However, Yau did the opposite and proved Calabi's conjecture in 1976. Thus such spaces are called Calabi–Yau spaces [Yau, 1977].

Yau's account of his struggle with the proof process can be found in [Yau and Nadis, 2010].

A Calabi–Yau space CY(n), also calld a CY(n-fold), must have several qualities:

1. CY(n) is a complex n-dimensional manifold.
2. CY(n) has structure group $\mathcal{SU}(n)$.
3. CY(n) has both a Kähler and complex metric.
4. CY(n) has a Ricci-flat Kähler metric.

A 10D superstring theory can be compactified on a CY(3), leaving four uncompactified dimensions corresponding to ordinary spacetime. Supersymmetry on CY(3) stabilizes it from collapse. However, it is assumed to be of submicroscopic size, perhaps as small as a few orders of magnitude larger than the Planck length. Therefore much work has been done on models of superstring theory, such as heterotic and IIA string theory, with CY(3) compactifications.

It was in this context that pairs of CY(3) with identical moduli spaces (and thus the same physical predictions) were discovered by [Greene and Plesser, 1990]. They called this CY(3) duality "mirror symmetry."

The CY(3) as a 6D real manifold can be described as a 3D sphere S^3 with a 3D torus T^3 attached at every point. Thus the strings living on the CY(3) can wind around a T^3 any number of times, called the winding number w. The string also has a quantized momentum m, which is dual to w in the sense that m on a T^3 with radius r is equivalent to w on T^3 with radius $1/r$.

It is this duality between r and $1/r$ that characterizes a dual pair of CY(3)s. Another aspect of this duality is that mirror pairs of CY(3)s must have opposite topologies in the sense of opposite Euler characteristic χ. Since χ of an ordinary T^3 is 0, these opposite χs must be due to the numbers (a and b) of T^3s with singularities each having $\chi = +1$ or -1.

The mirror duality implies opposite topologies:

$$a(+1) + b(-1) = \chi \text{ in one CY(3)},$$
$$a(-1) + b(+1) = -\chi \text{ in the dual CY(3)}.$$

This geometrical picture is a consequence of the conjecture by [Strominger *et al.*, 1996], with the title, "Mirror Symmetry is T-duality." Since T-duality transforms between r and $1/r$, in the T^3s, it is appropriate that T stands for Torus.

Although this SYZ conjecture has acquired increasing support in string theory applications, it has yet to be proved [Becker *et al.*, 2007].

It has been estimated that there are at least 10^{500} mirror pairs of CY(3)s, cf. [Yau and Nadis, 2010]. Moreover, as they mention, the singular ("bad") Tori are located at points of intersection of a network on the S^3 moduli space. Thus these mirror pairs would correspond to different configurations of points on these S^3 networks.

Here I will make the (perhaps too bold) conjecture that there must be an ADE classification of these mirror pairs, which would imply an infinity of such pairs, but they would be well organized.

This conjecture depends on the fact that the S^3 can be considered as equivalent to the Lie group $\mathcal{SU}(2)$. Then each McKay group as a finite subgroup of $\mathcal{SU}(2)$ is a set of points along a network on S^3. My conjecture is that each such ADE classified set of points corresponds to a CY(3) mirror pair, cf. Chap. 4.

An indirect support for this ADE classification of CY(3)s is the fact that there already is an ADE classification of CY(2)s.

This CY(2) ADE classification is based on the structure:

$$\text{CY}(2) = T^2 \times S^2 \text{ analogous to } \text{CY}(3) = T^3 \times S^3.$$

Moreover, the relationship between S^3 and S^2 is the Hopf fibration:

$$S^3 \to S^2 \quad \text{with } S^1 \text{as fiber.}$$

Thus with the parallel projection ($T^3 \to T^2$), it is plausible to consider the lifting:

$$T^2 \times S^2 \quad \text{to } T^3 \times S^3$$

so that CY(2) embeds in CY(3).

There are only 2 CY(2)s: T^4 and K3 spaces (which were named for Kummer, Kähler and Kodaira). And if we require $\mathcal{SU}(2)$ as structure (holonomy) group, only K3 is considered as CY(2). Although there is an infinity of K3 spaces, they are all diffeomorphic to each other, and thus are topologically equivalent. In a parallel to the CY(3) case, a K3 space can be described as an elliptic fibration of the sphere S^2. That is:

$$K3 \to S^2 \quad \text{with } T^2 \text{ as fiber.}$$

Also, as in the CY(3) case, the infinity of different K3s corresponds to the pattern of singular T^2 fibers living on S^2. In this case there are always 24 singular T^2 fibers. Thus, unlike the CY(3) case, T-duality changes the K3 into itself, so that K3 in this sense is its own mirror.

Moreover, the 24 singular fibers correspond to the Euler characteristic $\chi(\text{K3})$ being 24 [Yau and Nadis, 2010].

It happens that there is a close relationship between K3 manifolds and the Asymptotically Locally Euclidean (ALE) spaces of Chap. 8.

This relationship can be described in a series of steps [Gimon and Johnson, 1996]:

1. Construct the orbifold limit of the K3 as $T^4/\mathbb{Z}_n$.
2. Notice that this limit is a manifold, which is smooth, except at points fixed by the action of $\mathbb{Z}_n$.
3. Locally the region around each fixed point is a copy of $\mathfrak{R}^4$.
4. Remove an S^3 region around the fixed point.
5. This S^3 can be considered as the asymptotic region of an ALE space, which is $\mathfrak{R}^4/\boldsymbol{m}$, where $\boldsymbol{m}$ is a McKay group (cf. Chap. 4).

This relationship between K3 and ALE spaces suggests that string theory compactifications can be modeled directly on ALE spaces (which are ADE classified).

Indeed, this strategy is described in great detail in [Johnson and Myers, 1997].

This strategy is based on the self-duality of the type **IIB** theory. This is the S (strong/weak) duality as described in Chap. 8.

There are many consequences of this ALE strategy. Perhaps the most striking is the fact that the open strings can be regarded as D1-branes in an ALE space which are projected down by the McKay group $\boldsymbol{m}$ to the singular point in $\mathfrak{R}^4/\boldsymbol{m}$. The bouquet of S^2 spheres collapsed to a point by this projection corresponds exactly to the structure of the gauge group defining the set of vector multiplets of particles. Let the following table make this clear.

Label	McKay group	Gauge group for vector multiplets
A_n	$\mathbb{Z}_{n+1}$	$\mathcal{U}(1)^{n+1}$
D_n	$\mathcal{Q}_{n-2}$	$\mathcal{U}(1)^4 \times \mathcal{U}(2)^{n-3}$
E_6	$\mathcal{TD}$	$\mathcal{U}(1)^3 \times \mathcal{U}(2)^3 \times \mathcal{U}(3)$
E_7	$\mathcal{OD}$	$\mathcal{U}(1)^2 \times \mathcal{U}(2)^3 \times \mathcal{U}(3)^2 \times \mathcal{U}(4)$
E_8	$\mathcal{ID}$	$\mathcal{U}(1) \times \mathcal{U}(2)^2 \times \mathcal{U}(3)^2 \times \mathcal{U}(4)^2 \times \mathcal{U}(5) \times \mathcal{U}(6)$

Cf. Chap. 8.

Note especially that the index numbers i for $\mathcal{U}(n)^i$ correspond exactly to the balance numbers of the extended Coxeter graphs.

Also note that the complex group algebra of the McKay group has embedded as the unitary subgroup exactly the gauge group in this table.

For example, in the E_7 case, the extended Coxeter graph (with balance numbers attached) is:

```
1—2—3—4—3—2—1
      |
      2
```

Thus the complex group algebra $\mathbb{C}[\mathcal{OD}]$ has embedded as unitary subgroup exactly the corresponding gauge group:

$$\mathcal{U}(1)^2 \times \mathcal{U}(2)^3 \times \mathcal{U}(3)^2 \times \mathcal{U}(4)$$

cf. Chap. 3.

Note especially that the Standard Model gauge group is, $\mathcal{U}(1) \times \mathcal{SU}(2) \times \mathcal{SU}(3)$, and is easily seen as a subgroup of the ALE generated gauge groups of E_6, E_7 or E_8.

In this ALE version of **IIB** string theory, there are also hypermultiplets of types **I** and **II**. This is also a consequence of the McKay group representation structure acting on different dimensional sectors of the 10D spacetime, $\mathfrak{R}^6 \times \mathfrak{R}^4/\boldsymbol{m}$, which we label with the indices {0, 1, 2, 3, 4, 5} and {6, 7, 8, 9}.

Vector Multiplets

{0,1} 2D space in $\mathfrak{R}^6$: D1-branes sweep out a 2D worldsheet with parameters {0,1}. These D1-branes live in the orbifold $\mathfrak{R}^4/\boldsymbol{m}$ in an $\boldsymbol{m}$-invariant way. This means that there are $|\boldsymbol{m}|$ such D1-branes occupying the singularity structure of $\mathfrak{R}^4/\boldsymbol{m}$ at the origin of $\mathfrak{R}^4$. This D1-worldsheet structure is invariant under a subgroup of the adjoint representation $\mathcal{U}(|\boldsymbol{m}|)$, where $|\boldsymbol{m}|$ is the order of the McKay group. This subgroup consists of a product of $\mathcal{U}(n)$ groups corresponding exactly to the structure of the Coxeter graph. This is the gauge group of the vector multiplets, and is of dimension equal to the sum of the fundamental representations of the constituent unitary groups. This sum is equal to the Coxeter number of the Coxeter graph.

$$\mathrm{E}_7 \text{ case: } \mathcal{U}(1)^2 \times \mathcal{U}(2)^3 \times \mathcal{U}(3)^2 \times \mathcal{U}(4),$$

with total dimensionality:

$2+6+6+4=18$, which is the E_7 Coxeter number.

Hypermultiplets I

$\{0,\ldots,5\}$ $\mathfrak{R}^6$ space: these are vectors acted on by the adjoint representation of the gauge group as subgroup of $\mathcal{U}(|m|)$, which is a representation of order $|m|$.

E_7 case: $2+12+18+16=48$. Note that 48 is the sum of the squares of the E_7 balance numbers: {1, 2, 3, 4, 3, 2, 1, 2}.

Hypermultiplets II

$\{6,\ldots,9\}$ $\mathfrak{R}^4/m$ space: describing the motion of the open strings stretching between and attaching to parallel D1-branes in $\mathfrak{R}^4/m$. Amazingly, the ADE extended Coxeter graphs also provide the structure of these hypermultiplets. This is encoded in the links between nodes of this graph, so that the sum of the dimensions is the order of the McKay group $|m|$.

E_7 case: 1—2—3—4—3—2—1
with a node 2 attached by a vertical link below the 4

This graph implies the multiplets:

$$\mathbf{(1,2)}+\mathbf{(2,3)}+\mathbf{(3,4)}+\mathbf{(3,4)}+\mathbf{(2,3)}+\mathbf{(1,2)}+\mathbf{(2,4)}$$

so that the total dimensional representation is:

$$2+6+12+12+6+2+8=48.$$

There are many other aspects of this **IIB** string theory on ALE spaces, which is described in [Johnson and Myers, 1997].

Moreover, the role of $\mathfrak{R}^4/m$ can be continued to the theme of ADE type gauge groups corresponding to Calabi–Yau type compactifications in CY(4) spaces.

In this case, CY(4) is the compactifaction structure of the F-theory, which is a 12-dimensional spacetime theory.

F-theory was discovered by [Vafa, 1996] as the strongly coupled version of the **IIB** string theory (which is ten-dimensional). This means that there is a strong–weak duality (called S-duality) between F-theory and **IIB** theory. This duality corresponds to the action of the $\mathcal{SL}(2, \mathrm{Z})$ group on T^2 which compactifies the 12 dimensions of F-theory down to ten uncompactified dimensions which are the ten spacetime dimensions of the **IIB** theory.

Here we will consider the CY(4) compactification of F-theory which has a rather fascinating ADE classification, cf. [Heckman, 2010].

F-theory is an open string theory, in which the strings end on D7-branes, which are seven-dimensional membranes with a Dirichlet boundary. A stack of D7-branes can fill an 8D spacetime. Since F-theory is a 12D spacetime theory, there is a rather intricate structure to the compactifaciton of these 12 dimensions by the 8D CY(4) space.

First we think of the CY(4) as $\mathbb{C}^2 \times \mathbb{C}^2/m$ (where m is a McKay group and thus is ADE classified). Then we can consider the 12D spacetime as:

$$\mathfrak{R}^{3,1} \times \mathbb{C}^2 \times \mathbb{C}^2/m = \mathfrak{R}^{3,1} \times \mathfrak{R}^4 \times \mathbb{C}^2/m = \mathfrak{R}^{7,1} \times \mathbb{C}^2/m.$$

Thus the D7-brane as a spacetime structure fills $\mathfrak{R}^{7,1}$.

Note that $\mathbb{C}^2 \times \mathbb{C}^2/m$ is not really compact, so that the compact version of the compacification space would be:

$$T^4 \times \mathrm{K3},$$

where this K3 is the ALE version of $\mathbb{C}^2/m$ (cf. Chap. 8). However, we should not consider F-theory as including gravity, since we are here in the realm of a strongly coupled theory near the Planck scale of energy. Accordingly, there are no closed strings in this theory, but rather open strings attached to the D7-brane.

As a consequence of all this F-theory structure, there is a rather magical relationship for the spacetime structure $\mathfrak{R}^{7,1} \times \mathbb{C}^2/m$.

For each ADE type of $\mathbb{C}^2/m$ there is the corresponding ADE type of gauge group acting on the D7-brane structure in $\mathfrak{R}^{7,1}$. This D7-brane structure corresponds exactly to the deformation structure of the corresponding ADE singularity. For example, in the E_7 case, there would be 7 D7-branes corresponding to the bouquet of 7 S^2 spheres which desingularize the $\mathbb{C}^2/\mathcal{OD}$ singularity (cf. Chap. 7):

OO$\underset{\text{O}}{\text{O}}$OOO

These relationships can be described by the following table:

Label	McKay group m	Gauge group	Dimension
A_n	$\mathcal{Z}_{n+1}$	$\mathcal{SU}(n+1)$	n^2+2n
D_n	$\mathcal{Q}_{n-2}$	$\mathcal{SO}(2n)$	$(4n^2-2n)/2$
E_6	$\mathcal{TD}$	E_6	78
E_7	$\mathcal{OD}$	E_7	133
E_8	$\mathcal{ID}$	E_8	248

The F-theory compactified on K3$\times T^2$ as a CY(4) compactifiction is Dual to **IIB** theory compactified on K3. This is an S-duality, in which F-theory is strongly coupled while **IIB** is weakly coupled.

Furthermore, F-theory compactified on K3 is dual to the heterotic theory compactified on T^2. Let the following table summarize this and other K3 dualities.

K3 compactifications in string theory dualities.

Theory	S-T dim.	Moduli space	Dim.	Feature
F-th. on K3	$8+4$	$\mathfrak{R}^+ \times \mathfrak{M}_{18,2}$	37	7-branes
Het. on T^2	$8+2$	$\mathfrak{R}^+ \times \mathfrak{M}_{18,2}$	37	$\mathfrak{R}^+$ coupling c.
M-th. on K3	$7+4$	$\mathfrak{R}^+ \times \mathfrak{M}_{19,3} \times \mathfrak{R}^{22}$	80	$\mathfrak{R}^{22}$ lattice
Het. on T3	$7+3$	$\mathfrak{R}^+ \times \mathfrak{M}_{19,3} \times \mathfrak{R}^{22}$	80	$\mathfrak{R}^{22}$ lattice
IIA on K3	$6+4$	$\mathfrak{R}^+ \times \mathfrak{M}_{20,4}$	81	$\mathfrak{R}^+$ dilaton
Het on K3	$6+4$	$\mathfrak{R}^+ \times \mathfrak{M}_{20,4}$	81	$\mathfrak{R}^+$ dilaton
IIB on K3 $\times$ S1	$5+5$	$\mathfrak{R}^+ \times \mathfrak{M}_{21,5}$	106	26 U(1)s
Het on T^5	$5+5$	$\mathfrak{R}^+ \times \mathfrak{M}_{21,5}$	106	26 U(1)s

Cf. [Becker *et al.*, 2007; Aspinwall, 1999; Fischler and Rajaraman, 1997].

The role of the ADE groups can be clarified in these structures by describing in more detail the structure of the moduli spaces. As an examples we can describe the **IIA** and M-th. moduli spaces.

81 dim. = 58 (K3 metric) + 22 (B field) $+\mathfrak{R}^+$ dilaton

$\mathfrak{R}^+ \times \mathfrak{M}_{20,4}$ where $\mathfrak{M}_{20,4}$ means:

$\mathcal{O}(4,20)/\mathcal{O}(4) \times \mathcal{O}(20)$ which corresponds to:

$$\mathrm{D}_{12}/(\mathrm{A}_1 \times \mathrm{A}_1) \times \mathrm{D}_{10},$$

where D_{12} is a 276-dimensional Lie group $\mathcal{SO}(24)$; $\mathrm{A}_1 \times \mathrm{A}_1$ is a 6-dimensional Lie group $\mathcal{SO}(4) = \mathcal{SU}(2) \times \mathcal{SU}(2)$; and D_{10} is a 190-dimensional Lie group $\mathcal{SO}(2)$.

Thus the $\mathcal{R}^+ \times \mathfrak{M}_{20,4}$ moduli space has dimensionality:

$$1 + 276 - 6 - 190 = 81.$$

Note that the dimensionality of $\mathfrak{M}_{20,4}$ is $20 \times 4 = 80$, which is implied by the Lie group structures.

M-theory compactified on K3 has the 80-dimensional moduli space $\mathfrak{R}^+ \times \mathfrak{M}_{19,3} \times \mathfrak{R}^{22}$, so that $1+(3\times19)+22 = 80$. In this case the moduli space of K3 itself is $\mathfrak{R}^+ \times \mathfrak{M}_{19,3}$, where the $58 (= 2(19)+1)$ dimensions. They consist of 19 complex structure moduli and 20 Kähler moduli. This sums to $58 = 2(19) + 20$.

The $\mathfrak{R}^{22}$ lattice corresponds to 22 U(1) gauge fields in the seven uncompactified dimensions. And since the desingularization of K3 corresponds to the ADE type bouquets of S^2 two-cycles, the ADE Coxeter graphs relevant for group action on the seven uncompactified dimensions come into the picture. On the one hand, the ADE Coxeter graphs correspond to the Cartan subgroup, which is a product of $\mathcal{U}(1)$s. On the other hand, the ADE Coxeter graphs correspond to the non-Abelian gauge symmetry described by the ADE Lie group.

Thus the $\mathfrak{R}^{22}$ lattice must correspond to a total sum of ranks of ADE graphs. For example:

$$\mathrm{E}_8 \times \mathrm{E}_8 \times \mathrm{A}_3 \times \mathrm{A}_3 \quad \text{or} \quad \mathrm{E}_7 \times \mathrm{E}_7 \times \mathrm{E}_7 \times \mathrm{A}_1.$$

Note that M-theory not only unifies the five competing 10D superstring theories, but has a low energy 11D point particle limit which is a supergravity theory in which E_7 is the preferred symmetry for a $4+7$ spacetime cf. Chap. 8.

For E_7 supegravity, see [de Wit and Nicolai, 1986].

Appendix: Umbral Moonshine on K3

In Chap. 12, the largest sporadic finite group $\mathfrak{M}$ and its "moonshine" relationship to the coefficients of the j-function was described.

Very recently an analogous relationship between the Mathieu 24 group M_{24} and an elliptic genus function on the K3 space. This relationship was proved by [Gannon, 2013].

Since then, 22 more such correspondences (called "Umbral Moonshine") have been proved by [Duncan *et al.*, 2015]. Moreover, all 23 of these group structures are directly related to the geometry and physics of K3, cf. [Cheng and Harrison, 2015].

The key to these correspondences lies in the fact that the singularity structure of K3 is ADE classified by the orbifolds $\mathbb{C}^2/\mathcal{m}$, where $\mathcal{m}$ is a finite subgroup of $\mathcal{SU}(2)$. The ADE classification of $\mathcal{m}$ implies that the corresponding ADE Coxeter graph depicts the bouquet of S^2 spheres which structure the series of blow-ups that desingularize the singular point at the origin of $\mathbb{C}^3$, which embeds the $\mathbb{C}^2/\mathcal{m}$ orbifold.

Now, as mentioned above, the Euler characteristic of K3 is $\chi = 24$; and this implies that there are 24 singular 2-Tori in the K3 structure $S^2 \times T^2$. Thus a direct sum of ADE-type Coxeter graphs (if the sum of their ranks is 24) can describe the desingularization of these 24 singular points of K3. For this to work, each part of the ADE sum must have the same Coxeter number, which is the sum of the balance numbers of the extended Coxeter graph (cf. Chap. 6).

These 23 ADE structures correspond exactly to the 23 Niemeier lattices, which are 24-dimensional even, unimodular lattices. The 24D Leech lattice is also an even, unimodular lattice. This is a very special lattice which is not describable in terms of ADE root structures.

Rather, as described in Chap. 23 of *Sphere Packings, Lattices and Groups* [Conway and Sloane, 1988], the 23 Niemeier lattices correspond to the 23 "deep holes" in the Leech lattice.

The Leech lattice has as automorphism group the Conway group CO_1 which is one of 26 sporadic finite groups, the largest of which is the Monster $\mathfrak{M}$ cf. [Ronan, 2006].

The Leech lattice is rather magical since it describes the densest sphere-packing in $\mathfrak{R}^{24}$ space. In this sphere-packing, each S^{23} sphere touches 196,560 others.

Most amazingly, however, both the Leech lattice and the 23 Niemeier lattices describing its 23 deep holes can be derived as sublattices of the even unimodular Lorentzian lattice $\mathrm{II}_{25,1}$ in $\mathfrak{R}^{25,1}$.

String theorists will readily recognize $\mathfrak{R}^{25,1}$ as the spacetime of the 26D bosonic string theory, and the bosonic part of the heterotic string theory. Moreover, since the string sweeps out a 2D worldsheet embedded in this 26D space, there are 24 transverse vibrational degrees of freedom on this 2D worldsheet. This 24D transverse vibrational structure enters into the description of the heterotic string theory by way of the length-24 Golay code, and the equivalent use of three copies of the Hamming-8 code described by the lattice structure of the $\mathrm{E}_8 \times \mathrm{E}_8 \times \mathrm{E}_8$ Coxeter graphs and the corresponding Lie algebras cf. [Frenkel *et al.*, 1984, 1985].

Now the automorphism group of the Golay-24 code is the Mathieu-24 group M_{24} whose lattice corresponds to A_1^{24}. The K3 singularity structure that corresponds to M_{24} entails an elliptic genus whose Fourier expansion has an infinite series of coefficients, beginning with:

$$45, 231, 770, 2277, 5796, 13915, 30843, 65550, 132825, \ldots.$$

These coefficients correspond to the 26 irreducible representation dimensions of M_{24}.

1, 23, 45, $\underline{45}$, 231, $\underline{231}$, 252, 253, 483, 770, $\underline{770}$, 990, $\underline{990}$, 1035, $\underline{1035}$, 1035, 1265, 1771, 2024, 2277, 3312, 3520, 5313, 5796, 5544, 10395 (where the underlined numbers refer to complex conjugate representations).

The K3 coefficients correspond directly to M_{24} irreducible dimensions or to a sum of such dimensions. For example:

$$13915 = 3520 + 10395,$$

$$30843 = 10395 + 5796 + 5313 + 2014 + 1771$$

cf. [Eguchi *et al.*, 2010].

This matching of elliptic genus coefficients with representations of M_{24} is analogous to the matching of the Fourier expansion coefficients of the j-function with the representation dimensions of the Monster group $\mathfrak{M}$. This observation led to the Monsterous Moonshine conjectures of [Conway and Norton, 1979] cf. [Conway and Sloane, 1988] and Chap. 12.

These conjectures were all proved by [Borcherds, 1992; 2002] cf. [Ronan, 2006].

Similar matchings of elliptic genus coefficients with group representations (called Umbral Moonshines) occur for all 22 of the other K3 ADE structured Niemeier lattices.

There are several ADE-type ingredients that contribute to the K3 elliptical genus.

(1) The Klein–Duval singularities corresponding to $\mathbb{C}^2/m$, where m is a finite subgroup of $\mathcal{SU}(2)$ cf. [Klein, 1956; Duval, 1964] see Chap. 7.
(2) The ADE matrices Ω^ϕ of Capelli–Itzykson–Zuber, which contribute the "mockiness" of the mock modular forms for the 23 cases of umbral moonshine. Here, Φ labels the ADE type root system.

Note that mock modular forms were first introduced by Shrinivasa Ramanujan as "mock theta functions" in his last letter to Hardy in 1920. This was closely related to Ramanujan's work on partitions, which are closely related to the elliptical genus functions [Klarreich, 2015; Andrews, 1976].

The ADE classification of these Ω^Φ matrices is rather intricate and can be summarized by the following table, cf. [Cheng and Harrison, 2015].

Φ	Ω^ϕ
A_{m-1}	$\Omega_m(1)$
$D_{m/2+1}$	$\Omega_m(1) + \Omega_{m(m/2)}$
E_6	$\Omega_{12}(1) + \Omega_{12}(4) + \Omega_{12}(6)$
E_7	$\Omega_{18}(1) + \Omega_{18}(6) + \Omega_{18}(9)$
E_8	$\Omega_{30}(1) + \Omega_{30}(6) + \Omega_{30}(10) + \Omega_{30}(15)$

Here, Φ is an ADE root system, and Ω^ϕ is a $2m \times 2m$ matrix, where m is the ADE Coxeter number.

(3) In order to describe the elliptic genus function for any of the 23 Umbral groups, it is necessary to enlarge the Ω matrix to a sum of Ω^{ϕ} matrices:

$$\Omega^X = \sum_i \Omega^{\phi}(i),$$

where X labels the umbral group corresponding to the Niemeier lattice L^X. Note that X is a union of ADE type root systems:

$$X = \bigcup_i \phi^i.$$

All these notations are clarified by the explicit listing of the 23 umbral groups G^X corresponding to the 23 Niemeier lattices L^X [Duncan *et al.*, 2015; Cheng and Harrison, 2015].

The 23 Niemeier lattices X and their umbral groups G^X.

X	G^X	X	G^X	X	G^X	X	G^X	X	G^X
A_1^{24}	M_{24}	A_6^4	$\mathcal{SL}_2(3)$	A_{12}^2	$\mathbb{Z}_4$	D_6^4	Sym_4	D_{24}	$\mathbb{Z}_1$
A_2^{12}	$2 \cdot M_{12}$	$A_7^2 D_5^2$	Dih_4	$A_{15}D_9$	$\mathbb{Z}_2$	D_8^3	Sym_3	E_6^4	$\mathrm{GL}_2(3)$
A_3^8	$2 \cdot \mathrm{AGL}_3(2)$	A_8^3	Dih_6	$A_{17}E_7$	$\mathbb{Z}_2$	$D_{10}E_7^2$	$\mathbb{Z}_2$	E_8^3	Sym_3
A_4^6	$\mathrm{GL}_2(5)/2$	$A_9^2 D_6$	$\mathbb{Z}_4$	A_{24}	$\mathbb{Z}_2$	D_{12}^2	$\mathbb{Z}_2$		
$A_5^4 D_4$	$\mathrm{GL}_2(3)$	$A_{11}D_7E_6$	$\mathbb{Z}_2$	D_4^6	$3 \cdot \mathrm{Sym}_6$	$D_{16}E_8$	$\mathbb{Z}_1$		

Note that $\mathrm{GL}_3(2) = \mathrm{PSL}_2(7)$, which is the automorphism group of the unique genus 3 curve (over $\mathbb{C}$), called the Klein quartic curve. $\mathrm{PSL}_2(7)$ is a simple group of order 168, which corresponds to the 168 shaded regions of Klein's quartic curve [Gray, 1982; Elkies, 1998].

$\mathrm{PSL}_2(7)$, of order $168 = 24(7)$, is also the automorphism group of the E_7 (Hamming 7) code, cf. [Conway and Sloane, 1988].

Note also that the E_7 quartic on $\mathbb{C}^2$ is $x^3 + xy^3$, whose zero set is the genus 3 Riemann surface, with 28 bitangents corresponding to the 56D fundamental representation of the E_7 Lie algebra [Hartshorne, 1977; Clemens, 1980].

Conjecture. *Since* $S_4 \subset \mathrm{PSL}_2(7)$, *the* E_7 *quartic curve* (*of genus* 3) *corresponds to the* 3*-family structure of* S_4 *cf. Chap.* 2.

Chapter 17

Heisenberg Algebras

The mathematical term, Heisenberg algebra, is based on the early history of quantum mechanics in which Heisenberg [1925], Born and Jordan [1925], Dirac [1926], Born, Heisenberg, and Jordan [1926] first published the idea that quantum mechanics fundamentally entailed the non-commutative algebra with the formula, cf. [Heisenberg, 1930; Van Der Waerden, 1968]:

$$PQ - QP = \hbar/2\pi i,$$

where the matrix P is the momentum operator acting on a vector space of states for a quantum particle, the matrix Q is the position operator; $\hbar$ is Planck's constant, and $i = \sqrt{-1}$.

From this non-commutativity of P and Q, the Heisenberg uncertainty principle can be derived [Martin, 1981]:

$$\Delta p \Delta q \geq \hbar/4\pi,$$

where p (momentum) and q (position) are based on the original 1D version of the Heisenberg algebra.

In physics, the three-dimensionality of ordinary space implies that P and Q are vectors in the symplectic space $\Re^3 + \Re^3$. Thus the original Heisenberg algebra is a 7D algebra, with the 7th dimension consisting of the Planck element $\hbar/2\pi$i cf. [Guillemin and Sternberg, 1984].

In mathematics, one can define a Heisenberg algebra for any symplectic space of dimension $\Re^{2n}$, since the vector-space $\Re^{2n}$ can be made into a Lie algebra by defining a Lie bracket on $\Re^{2n}$. Thus we would have:

$$[P_i, Q_j] = 0 \quad \text{where } \{i, j\} \text{ run through } \{1, \ldots, n\}.$$

This is a commutative Lie algebra, which is structurally equivalent to the Poisson algebra of classical mechanics.

As Paul Dirac has emphasized, the transition from classical to quantum physics corresponds to the transition from the commutative Poisson algebra to the non-commutative extension of this algebra [Dirac, 1926; 1958].

Physicists consider the Schrödinger wave equation as an alternative version (called a "picture") of quantum mechanics. So we have the Heisenberg picture and the Schrödinger picture, as well as the independently developed Dirac transformation version of quantum mechanics, which incorporates aspects of both of the Heisenberg and Schrödinger pictures [Schiff, 1968].

In order to show the equivalence between the Heisenberg and Schrödinger versions of quantum mechanics, Herman Weyl derived the Schrödinger wave function from the Heisenberg algebra [Weyl, 1950].

Mathematicians call the extension from the commutative Poisson algebra to the Heisenberg algebra a central extension, because the extra element is in the center of the algebra. In fact, by setting $\hbar/2\pi i$ equal to I as the identity matrix, the Heisenberg algebra is defined by way of the commutator (Lie bracket) formulas:

$$[P_i, Q_j] = I,$$
$$[P_i, P_j] = 0,$$
$$[Q_i, Q_j] = 0,$$

where i and j range through integers from 1 to n for the $\Re^{2n}$ symplectic space.

Thus the Heisenberg algebras are Lie algebras. As we will see they are Lie subalgebras of simple Lie algebras. In fact, as [Kostant, 1984; 1985; 2004] has shown, these Heisenberg algebras are subalgebras of the ADE type Lie algebras cf. Chap. 5.

Kostant uses the ADE classification of the McKay correspondence between the affine Kac–Moody Lie algebras (which are infinite dimensional) and the finite subgroups of $\mathcal{SU}(2)$. cf. Chap. 4.

Thus there is a bijection between the finite subgroups of $\mathcal{SU}(2)$, which I call McKay groups $\mathfrak{m}$, and the finite dimensional (ADE classified) Lie algebras $\mathcal{L}$. This is because $\mathcal{L}$ is naturally a Lie subalgebra of the affine Kac–Moody Lie algebra, ${}^{\infty}\mathcal{L}$.

Thus in the full generating functions:

$$P_m(t)_i = \sum_{j=\{0,\ldots,K\}} Z_j t^j/(1-t^a)(1-t^b),$$

$P_m(t)_0$ corresponds to $^\infty\mathcal{L}$, while $P_m(t)_{1\ldots,n}$ correspond to $\mathcal{L}$, where n is the ADE rank; K is the Coxeter number; $a+b=K+2$; and $ab = 2|m|$ cf. [Kostant, 2004; Stekolshchik, 2006].

[Kostant, 1984] displays the ADE consequences of the $P_m(t)$ generating function as follows (with my notation as in Chap. 4):

ADE classification: Lie algebra $\mathcal{L}$, McKay group m and K, a, b.

Lie algebra $\mathcal{L}$	McKay group m	$\lvert m \rvert$	a	b	K
A_{2n-1}	$\mathbb{Z}_n$	n	2	$2n$	$2n$
D_n	$\mathbb{Q}_n$	$4(n-2)$	4	$2(n-2)$	$2n-2$
E_6	$\mathcal{TD}$	24	6	8	12
E_7	$\mathcal{OD}$	48	8	12	18
E_8	$\mathcal{ID}$	120	12	20	30

Note that a, b and K are also the degrees of the invariants of the ADE catastrophe germs cf. Chap. 7.

Also note that the Coxeter number K is the sum of balance numbers on the extended Coxeter graph, which corresponds to the infinite order Coxeter reflection group as well as the affine Kac–Moody Lie algebra. Moreover, for D and E type graphs, $a = 2d$, where d is the balance number at the branch node, which is attached to three other nodes.

For example, the extended E_7 Coxeter graph (with $K = 18$) is:

```
1*—2—3—4—3—2—1
       |
       2
```

where {2, 3, 4, 3, 2, 1, 2} are coefficients of the highest root in the 7D reflection space $\mathfrak{R}^7$ of E_7. And we note that $a = 2(4)$.

Moreover, since $\{a, b, K\}$ are degrees of the invariants of the catastrophe germ, we have the E_7 germ:

$$X^3 + XY^3 + Z^2$$

$X = f^2$: deg. 12 invariant, where $f = xy^5 - x^5y$.

$Y = h$: deg. 8 invariant, where $h = x^8 + 14x^4y^4 + y^8$.
$Z = ft$: deg. 18 invariant, where $t = x^{12} - 33x^8y^4 - 33x^4y^8 + y^{12}$.

Felix Klein considered the 0-set of f as the six vertices of the octahedron; the 0-set of h as the eight vertices of the cube (dual to the octahedron); and the 0-set of t as the mid-edge points of the 12 edges of the octahedron [Klein, 1956].

A key aspect of Kostant's description of the ADE Heisenberg algebras is the use of the Coxeter element acting on the roots of the corresponding Lie algebra.

The Coxeter element w of the Coxeter group W is the product of the generating reflections of W. The ordering of these reflections is arbitrary, since all these orderings are conjugate to each other. For any element a of a finite group G, we say that $a^r = 1$ implies that $r = |a|$ the order of element a in G. Thus for the Coxeter element w of the Coxeter group W, it happens that:

$$|w| = K,$$

where K is the Coxeter number (as described above).

Note that [Coxeter, 1973] described the element w of W that came to be called the "Coxeter element" on pp. 233–234 of Chap. XII. This refers to a projection of an n-dimensional polytope onto a two-dimensional plane. This plane is called the Coxeter plane P, because it is defined by means of the Coxeter element w and its Coxeter number K.

The primitive roots of unity of order K correspond to rotations of P by w through an angle of $2\pi/K$.

This can be illustrated by the E_8 polytope projection, drawn by Peter McMullen as the frontispiece to [Coxeter, 1991]. More easily available is [Stembridge, 2007] which depicts projections to P of the polytopes, D_5, E_6, E_7 and E_8.

In these polytope projections to the Coxeter plane P, D_5 has an outer ring of eight nodes (and thus eight segments of a circle on P) corresponding to $2\pi/8$. There are four rings consisting of eight nodes each inside the outer ring, so that there is a total of $5(8) = 40$ nodes in the D_5 polytope.

Similarly, the polytope projections for E_6, E_7 and E_8, have outer rings of 12, 18 and 30 nodes, which correspond to $K = 12, 18$ and 30. Each iteration of the Coxeter element orbits the ring of nodes through an angle of $2\pi/12$, $2\pi/18$ and $2\pi/30$, respectively.

The fact that the rings of nodes correspond to orbits of the Coxeter element w (of the Coxeter group W) makes possible the projection of the polytope to the Coxeter plane P.

In each polytope, the number of vertices (nodes) is equivalent to the number of roots of the corresponding Lie algebra. Each root, as a vector in the reflection space, carries the eigenvalues of the non-commutative basis elements of the Lie algebra.

For an ADE type Lie algebra, with a set of roots Δ, the number of roots $|\Delta| = nK$, where n is the rank and K is the Coxeter number. The Lie algebra also has a commutative subalgebra $\mathfrak{h} = \mathfrak{R}^n$, called the Cartan subalgebra, so that the total dimensionality of the Lie algebra is $nK + n$.

Note that the reflection space is the dual space to the Cartan subalgebra $\mathfrak{h}$, so that the roots are located at the vertices of the (Coxeter) polytope in the reflection space $\mathfrak{R}^n$. Each root is orthogonal to its own reflection hyperplane (a mirror of dimension $n-1$); and each mirror has attached two roots on opposite sides of the mirror. Thus for nK roots there are $nK/2$ mirrors cf. Chap. 5.

The highest root ψ in Δ has a set of n coefficients, which are the n balance numbers (excluding the unit coefficient on the extended Coxeter graph). The highest root is:

$$\psi = \sum_{i=\{1,\ldots,n\}} d_i \alpha_i,$$

where n is the rank, and the coefficients d_i are n balance numbers and also the irreducible representation dimensions of the Coxeter group W. Moreover, the α_i are the n simple positive roots in the root system Δ of cardinality nK in the reflection space $\mathfrak{R}^n$.

For example, the coefficients of ψ for the E-type graphs are:

$$\psi(E_6):\{1,2,3,2,1,2\},$$
$$\psi(E_7):\{2,3,4,3,2,1,2\},$$
$$\psi(E_8):\{2,4,6,5,4,3,2,3\}.$$

Rather magically, ψ corresponds to a subset $\Phi \in \Delta$ of cardinality $2K-3$. Since the roots Φ correspond to a subset of Lie algebra basis elements, this subset constitutes a Heisenberg subalgebra of the Lie algebra. By viewing $2K-3$ as $2(K-2)+1$, we see that this Heisenberg subalgebra has $K-2$ momentum type operators and $K-2$ position type operators acting on a $K-2$ dimensional symplectic space; the extra term is, of course, the Planck element, and can by convention be the identity operator I.

The Coxeter element w transforms the highest root ψ through an orbit of K iterations of w. Calling each iteration:

$$\tau^{(x)} \quad \text{where } x = \{0, \dots, K-1\},$$

we can then construct the assembling vectors:

$$z_x = \tau^{(x-1)} - \tau^{(x)}$$

cf. [Stekolshchik, 2006].

For example, the E_6 Coxeter element $w(\mathrm{E}_6)$ orbits the highest weight vector $\tau^{(0)} = (1,2,3,2,1,2)$ through the iterations:

$$\begin{array}{ll}
\tau^{(1)} = (1,2,3,2,1,1), & \tau^{(2)} = (1,2,2,2,1,1), \\
\tau^{(3)} = (1,1,2,1,1,1), & \tau^{(4)} = (0,1,1,1,0,1), \\
\tau^{(5)} = (0,0,1,0,0,0), & \tau^{(6)} = (0,0,-1,0,0,0), \\
\tau^{(7)} = (0,-1,-1,-1,0,-1), & \tau^{(8)} = (-1,-1,-2,-1,-1,-1), \\
\tau^{(9)} = (-1,-2,-2,-2,-1,-1), & \tau^{(10)} = (-1,-2,-3,-2,-1,-1), \\
\tau^{(11)} = (-1,-2,-3,-2,-1,-2). &
\end{array}$$

Then the assembling vectors $z_x = \tau^{(x-1)} - \tau^{(x)}$ can be listed as:

$$\begin{array}{ll}
\mathcal{Z}_1 = (0,0,0,0,0,1), & \mathcal{Z}_2 = (0,0,1,0,0,0), \\
\mathcal{Z}_3 = (0,1,0,1,0,0), & \mathcal{Z}_4 = (1,0,1,0,1,0), \\
\mathcal{Z}_5 = (0,1,0,1,0,1), & \mathcal{Z}_6 = (0,0,2,0,0,0), \\
\mathcal{Z}_7 = (0,1,0,1,0,1), & \mathcal{Z}_8 = (1,0,1,0,1,0), \\
\mathcal{Z}_9 = (0,1,0,1,0,0), & \mathcal{Z}_{10} = (0,0,1,0,0,0), \\
\mathcal{Z}_{11} = (0,0,0,0,0,1), & \mathcal{Z}_{12} = (0,0,0,0,0,0).
\end{array}$$

By displaying these assembling vectors $\mathcal{Z}_x$ vertically, we have the following E_6 array:

$$\begin{array}{l} \quad\mathcal{Z}_1 \ldots\ldots\ldots\ldots\ldots \mathcal{Z}_{12} \\ 1_1\colon 0\,0\,0\,1\,0\,0\,0\,1\,0\,0\,0\,0 \\ 2_2\colon 0\,0\,1\,0\,1\,0\,1\,0\,1\,0\,0\,0 \\ 3_3\colon 0\,1\,0\,1\,0\,2\,0\,1\,0\,1\,0\,0 \\ 2_4\colon 0\,0\,1\,0\,1\,0\,1\,0\,1\,0\,0\,0 \\ 1_5\colon 0\,0\,0\,1\,0\,0\,0\,1\,0\,0\,0\,0 \\ 2_6\colon 1\,0\,0\,0\,1\,0\,1\,0\,0\,0\,1\,0 \end{array}$$

where the E_6 Coxeter graph with balance numbers indexed is:

$$\begin{array}{c} 1_1 - 2_2 - 3_3 - 2_4 - 1_5 \\ | \\ 2_6 \end{array}$$

From the pattern of 1's in this E_6 array we can write the polynomials: $\mathcal{Z}(t)_i$, where $i = \{1, \ldots, 6\}$:

$$\mathcal{Z}(t)_i = \sum_{j=\{0,\ldots,12\}} \mathcal{Z}_j t^j,$$

$$\begin{aligned} \mathcal{Z}(t)_1 &= t^4 + t^8, \\ \mathcal{Z}(t)_2 &= t^3 + t^5 + t^7 + t^9, \\ \mathcal{Z}(t)_3 &= t^2 + t^4 + 2t^6 + t^8 + t^{10}, \\ \mathcal{Z}(t)_4 &= t^3 + t^5 + t^7 + t^9, \\ \mathcal{Z}(t)_5 &= t^4 + t^8, \\ \mathcal{Z}(t)_6 &= t + t^5 + t^7 + t^{11}. \end{aligned}$$

Thus the generating functions for $\mathcal{TD}$, with $i = \{1, \ldots, 6\}$, are

$$P_{\mathcal{TD}}(t)_i = \sum_{j=\{0,\ldots,12\}} Z_j t^j / (1 - t^6)(1 - t^8).$$

Similarly, the seven generating functions for $\mathcal{OD}$ would be:

$$P_{\mathcal{OD}}(t)_i = \sum_{j=\{0,\ldots,18\}} Z_j t^j / (1 - t^8)(1 - t^{12}).$$

This corresponds to the E_7 Coxeter graph (with balance and indexing numbers):

$$\begin{array}{c} 2_1\text{–}3_2\text{–}4_3\text{–}3_4\text{–}2_5\text{–}1_6 \\ | \\ 2_7 \end{array}$$

The polynomials $\mathcal{Z}_j t^j$ (in accord with this indexing) are:

$$\begin{aligned}
\mathcal{Z}(t)_1 &= t + t^7 + t^{11} + t^{17},\\
\mathcal{Z}(t)_2 &= t^2 + t^6 + t^8 + t^{10} + t^{12} + t^{16},\\
\mathcal{Z}(t)_3 &= t^3 + t^5 + t^7 + 2t^9 + t^{11} + t^{13} + t^{15},\\
\mathcal{Z}(t)_4 &= t^4 + t^6 + t^8 + t^{10} + t^{12} + t^{14},\\
\mathcal{Z}(t)_5 &= t^5 + t^7 + t^{11} + t^{13},\\
\mathcal{Z}(t)_6 &= t^6 + t^{12},\\
\mathcal{Z}(t)_7 &= t^4 + t^8 + t^{10} + t^{14}.
\end{aligned}$$

Thus these polynomials would be derived from the E_7 array:

$$\begin{array}{l}
\qquad\mathcal{Z}_1 \ \ldots\ldots\ldots\ldots\ldots\ldots\ldots\ldots \ \mathcal{Z}_{18}\\
2_1\colon 1\,0\,0\,0\,0\,0\,1\,0\,0\,0\,1\,0\,0\,0\,0\,0\,1\,0\\
3_2\colon 0\,1\,0\,0\,0\,1\,0\,1\,0\,1\,0\,1\,0\,0\,0\,1\,0\,0\\
4_3\colon 0\,0\,1\,0\,1\,0\,1\,0\,2\,0\,1\,0\,1\,0\,1\,0\,0\,0\\
3_4\colon 0\,0\,0\,1\,0\,1\,0\,1\,0\,1\,0\,1\,0\,1\,0\,0\,0\,0\\
2_5\colon 0\,0\,0\,0\,1\,0\,1\,0\,0\,0\,1\,0\,1\,0\,0\,0\,0\,0\\
1_6\colon 0\,0\,0\,0\,0\,1\,0\,0\,0\,0\,0\,1\,0\,0\,0\,0\,0\,0\\
2_7\colon 0\,0\,0\,1\,0\,0\,0\,1\,0\,1\,0\,0\,0\,1\,0\,0\,0\,0
\end{array}$$

Moreover, as described above for the E_6 case, this array is derivable from the action of the Coxeter element $w(E_7)$ on the highest weight vector $\psi(E_7) = (2,3,4,3,2,1,2)$ in the E_7 Reflection space $\mathfrak{R}^7$.

Likewise, we can display the eight generating functions for the McKay group, $\mathcal{ID}$:

$$P_{\mathcal{ID}}(t)_i = \sum_{j=\{0,\ldots,30\}} Z_j t^j/(1-t^{12})(1-t^{20}).$$

This corresponds to the E_8 Coxeter graph, with balance and indexing nmbers:

$$\begin{array}{c} 2_1 - 4_2 - 6_3 - 5_4 - 4_5 - 3_6 - 2_7 \\ \qquad\quad | \qquad\qquad\qquad\qquad\qquad \\ \qquad\quad 3_8 \qquad\qquad\qquad\qquad\qquad \end{array}$$

$$\begin{aligned}
\mathcal{Z}(t)_1 &= t^7 + t^{13} + t^{17} + t^{23},\\
\mathcal{Z}(t)_2 &= t^6 + t^8 + t^{12} + t^{14} + t^{16} + t^{18} + t^{22} + t^{24},\\
\mathcal{Z}(t)_3 &= t^5 + t^7 + t^9 + t^{11} + t^{13} + 2t^{15} + t^{17} + t^{19} + t^{21} + t^{23} + t^{25},\\
\mathcal{Z}(t)_4 &= t^4 + t^8 + t^{10} + t^{12} + t^{14} + t^{16} + t^{18} + t^{20} + t^{22} + t^{26},\\
\mathcal{Z}(t)_5 &= t^3 + t^9 + t^{11} + t^{13} + t^{17} + t^{19} + t^{21} + t^{27},\\
\mathcal{Z}(t)_6 &= t^2 + t^{10} + t^{12} + t^{18} + t^{20} + t^{28},\\
\mathcal{Z}(t)_7 &= t + t^{11} + t^{19} + t^{29},\\
\mathcal{Z}(t)_8 &= t^6 + t^{10} + t^{14} + t^{16} + t^{20} + t^{24}.
\end{aligned}$$

The corresponding E_8 array is:

$\mathcal{Z}_1$ .... $\mathcal{Z}_{30}$

2_1: 0 0 0 0 0 0 1 0 0 0 0 0 1 0 0 0 1 0 0 0 0 0 1 0 0 0 0 0 0 0
4_2: 0 0 0 0 0 1 0 1 0 0 0 1 0 1 0 1 0 1 0 0 0 1 0 1 0 0 0 0 0 0
6_3: 0 0 0 0 1 0 1 0 1 0 1 0 1 0 2 0 1 0 1 0 1 0 1 0 1 0 0 0 0 0
5_4: 0 0 0 1 0 0 0 1 0 1 0 1 0 1 0 1 0 1 0 1 0 1 0 0 0 1 0 0 0 0
4_5: 0 0 1 0 0 0 0 0 1 0 1 0 1 0 0 0 1 0 1 0 1 0 0 0 0 0 1 0 0 0
3_6: 0 1 0 0 0 0 0 0 0 1 0 1 0 0 0 0 0 1 0 1 0 0 0 0 0 0 0 1 0 0
2_7: 1 0 0 0 0 0 0 0 0 0 1 0 0 0 0 0 0 0 1 0 0 0 0 0 0 0 0 0 1 0
3_8: 0 0 0 0 0 1 0 0 0 1 0 0 0 1 0 1 0 0 0 1 0 0 0 1 0 0 0 0 0 0

Remark I. For all three (E_6, E_7, E_8) structures depicted above, we notice that the pattern of 1's (and the single 2) indicates the pattern of the Heisenberg system Φ within the Lie algebra roots Δ. This pattern has three segments, which correspond to the Heisenberg algebra structure: $[P_i, Q_j] = I$.

$\Phi(P_i)$-left side; Φ(I)-central column; $\Phi(Q_i)$-right side.

Note that the 1's on the left side mirror the 1's on the right side.

Note also that the central column contains only the 2, which corresponds to the identity element I.

The same tri-partite pattern would also hold for all D-type Heisenberg algebras.

However, for the A-type Heisenberg algebras, these are defined only for odd rank Coxeter graphs: A_{2n-1}.

Such odd rank graphs have a central node, so that there is tri-partite structure $(L + 1 + R)$ for the Coxeter graph and thus for the elements of Φ.

Note that this excludes A-type graphs of even rank cf. [Kostant, 1984, 1985, 2004].

Remark II. First we note that the Heisenberg subalgebras of Lie algebras are classified by the finite subgroups of $\mathcal{SU}(2)$, which we call McKay groups. Moreover, the McKay correspondence depends on the structure of the affine Kac–Moody Lie algebras, which are infinite-dimensional Lie algebras cf. Chaps. 4 and 5.

Thus it is interesting that for affine Lie algebras of types $^{\infty}$A, $^{\infty}$D, and $^{\infty}$E, the principal subalgebra is an infinite-dimensional Heisenberg algebra cf. [Kac, 1985; Lepowsky and Wilson, 1985].

Moreover, the affine Coxeter graph has an extra node, which we can call the affine vertex, so that the full generating functions,

$$P_m(t)_i = \sum_{j=\{0,\ldots,K\}} Z_j t^j/(1-t^a)(1-t^b)$$

include the term for $i = 0$, as well as $i = \{1, \ldots, n\}, n =$ rank. ß.

For this affine vertex term [Kostant, 1984, 2004] has shown that the formula is very simple, and we can write:

$$P_m(t)_0 = 1 + t^K/(1-t^a)(1-t^b),$$

where K is the Coxeter number and $\{a, b\}$ are as in the table on p. 203.

For example: in the E_7 case, $K = 18, a = 8, b = 12$.

Chapter 18

Summary and Outlook

In 1972 Freeman Dyson cited an example in which "two disparate or incompatible mathematical concepts were juxtaposed in the description of a single situation." He urged mathematicians and physicists to "create a wider conceptual framework within which the pair of disparate elements would find a harmonious coexistence." [Lepowsky *et al.*, 1985].

In this book, we have explored many of the contours of the very wide conceptual framework of the ADE classifications. We have described many relationships between the mathematical (and physical) ADE classified objects.

The unifying theme has been the ADE Coxeter graphs, which are equivalent to the singly-laced Dynkin diagrams. This equivalence is due to the relationship between Coxeter's reflection groups and the so-called Weyl groups of the ADE Lie algebras cf. Chap. 6.

This equivalence is generalized by the equivalence between Coxeter's affine ADE diagrams and the ADE affine Kac–Moody Lie algebras. Here the infinite order Coxeter reflection groups correspond to the infinite-dimensional K-M algebras.

The K-M algebras become especially important in the McKay correspondence, which entails an ADE relationship between the affine K-M algebras and the finite subgroups of $\mathcal{SU}(2)$. Due to the importance of this correspondence, we have called these $\mathcal{SU}(2)$ subgroups McKay groups cf. Chap. 5.

The most useful summarizing table would be the affine Coxeter graphs as extended Coxeter graphs (with their balance numbers) and also the corresponding McKay groups, the finite subgroups of $\mathcal{SU}(2)$, with their orders indicated.

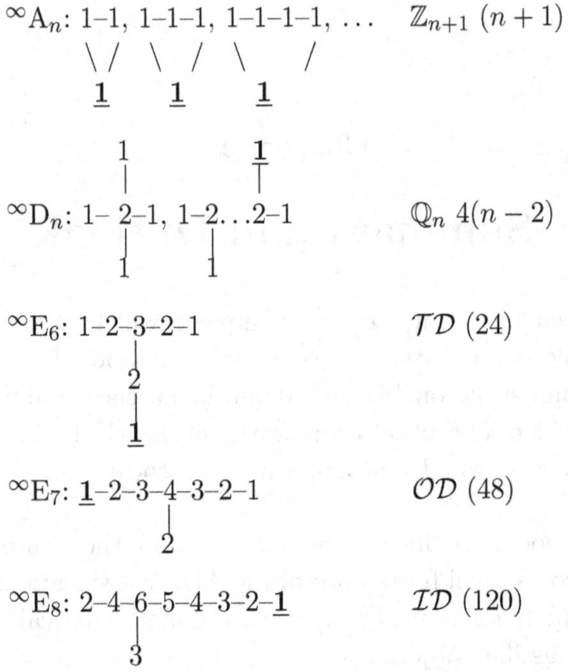

$^\infty A_n$: 1–1, 1–1–1, 1–1–1–1, ... $\quad \mathbb{Z}_{n+1}$ $(n+1)$

\/ \ / \ /

$\underline{1}$ $\underline{1}$ $\underline{1}$

1 $\underline{1}$

$^\infty D_n$: 1– 2–1, 1–2...2–1 $\quad \mathbb{Q}_n$ $4(n-2)$

1 1

$^\infty E_6$: 1–2–3–2–1 $\quad \mathcal{TD}$ (24)

2

$\underline{1}$

$^\infty E_7$: $\underline{1}$–2–3–4–3–2–1 $\quad \mathcal{OD}$ (48)

2

$^\infty E_8$: 2–4–6–5–4–3–2–$\underline{1}$ $\quad \mathcal{ID}$ (120)

3

Here the extended node is indicated by an underlined balance number. Thus the ADE Coxeter graph for finite reflection groups is simply the graph without the extended node. The remaining balance numbers are the coefficients of the highest weight in the Coxeter lattice, cf. Chap. 17.

There are many other properties of the ADE balance numbers. Here we list a few of them:

(1) The sum is the Coxeter number K. For example:

$$K(E_7) = 1 + 2 + 3 + 4 + 3 + 2 + 1 + 2 = 18.$$

(2) The sum of the squares of the balance numbers is the order of the McKay group $|\boldsymbol{m}|$. For example:

$$|\boldsymbol{m}|(E_7) = 1^2 + 2^2 + 3^2 + 4^2 + 3^2 + 2^2 + 1^2 + 2^2 = 48.$$

(3) The product of the balance numbers is a factor F of the Coxeter group order, which is $F(n!)C$, where n is the rank and C, is the

order of the center. Thus, for example, the order of the E_7 Coxeter group is:

$$288(5040)(2) = 2{,}903{,}040.$$

Note that for graphs of types D and E, the order of the center C is simply the number of 1's in the extended graph.

Note also, that for the A_n graphs, the Coxeter group is the symmetric group S_{n+1} of order $(n+1)!$.

(4) The balance numbers are the dimensions of the irreducible representations of the McKay group. This corresponds to the characters of the identity element in the McKay group character table. For example, in the E_7 case, the McKay group is $\mathcal{OD}$, whose identity characters are:

$$\{1, 2, 3, 4, 3, 2, 1, 2\}.$$

(5) The balance numbers record the dimensions of the total matrix algebras, which span the McKay group algebra. This accords with (2) above. This direct sum of matrix algebras have embedded a product of unitary groups $\mathcal{U}(n)$ such that the n's correspond exactly to the balance numbers.

For example in the E_7 case:

$\mathbb{C}[\mathcal{OD}]$ has embedded as the complete set of unitary elements:

$$\mathcal{U}(1) \times \mathcal{U}(2) \times \mathcal{U}(3) \times \mathcal{U}(4) \times \mathcal{U}(3) \times \mathcal{U}(2) \times \mathcal{U}(1) \times \mathcal{U}(2).$$

Since $\mathcal{U}(n) = \mathcal{U}(1) \times \mathcal{SU}(n)$, we have the rearrangement:

$$T^7 \times \mathcal{U}(1) \times \mathcal{SU}(2) \times \mathcal{SU}(3) \times \mathcal{SU}(4) \times \mathcal{SU}(3) \times \mathcal{SU}(2) \times \mathcal{SU}(2).$$

(6) The dimensionality d of the ADE Lie algebra is derivable from the rank n and the Coxeter number K according to the simple formula, cf. (1) above:

$$d = nK + n,$$

where nK corresponds to the non-commutative basis, while n corresponds to the dimensionality of the Cartan subalgebra, which is the largest commutative subalgebra.

Note also that nK is the number of (non-zero) roots in the ADE type Coxeter lattice. Thus $nK/2$ is the number of reflection planes acted on by the Coxeter group, which is generated by reflections in the n basic reflection planes, cf. Chaps. 5 and 6.

In the case of the E_7 lattice, we have:

$$nK/2 = 7(18)/2 = 63 \text{ reflection planes}$$

generated by the E_7 Coxeter group, of order 288(7!)(2), acting on the 7 basic planes cf. (3) above.

(7) Coxeter invariants and Coxeter exponents are derivable from the Coxeter element w of the Coxeter group W. The Coxeter element w corresponds to the highest root in the Coxeter lattice. The coefficients of the highest root are simply the balance numbers (excluding the balance number 1 at the extended node).

In the E_7 case the highest root coefficients are:

$$\{2, 3, 4, 3, 2, 1, 2\}.$$

From these coefficients of $w(E_7)$ and various combinatorial procedures, one can derive the E_7 Coxeter invariants of degree d_i and the exponents, $m_i = d_i - 1$, Thus we have:

E_7 Invariant degrees, $d_i : 2, 6, 8, 10, 12, 14, 18$.
E_7 Exponents, $m_i : 1, 5, 7, 9, 11, 13, 17$.

There are many other correspondences, including:

The sum of the E_7 exponents is 63, the number of reflecting planes (i.e. the mirror number).

The product of the E_7 Invariant degrees is 2,903,040, the order of the Coxeter group $w(E_7)$.

Note that the exponents generate the geometrical structure of the compact ADE Lie group. For example the compact E_7 Lie group (a 133D manifold) is the direct product of seven spheres, each of dimension $2(m_i) + 1$:

$$S^3 \times S^{11} \times S^{15} \times S^{19} \times S^{23} \times S^{27} \times S^{35}$$

where $3 + 11 + 15 + 19 + 23 + 27 + 35 = 133$

[Coxeter and Moser, 1965; Hiller, 1982; Kostant, 1984].

Moreover, 18 is the E_7 Coxeter number K, which is the sum of the E_7 balance numbers (in the extended Coxeter graph). Note that

$\{8, 12, 18\}$ are the degrees of the Klein invariants corresponding to the germ of the E_7 catastrophe cf. Chap. 7.

The same numbers $\{8, 12, 18\}$ are exponents in the Poincaré polynomial generating the infinite E_7 Heisenberg algebra:

$$P_{\mathcal{OD}}(t)_0 = 1 + t^{18}/(1 \times t^8)(1 \times t^{12})$$

cf. Chaps. 11 and 17.

(8) The ADE classification of ALE spaces has many consequences. For example, the ALE space is a desingularization of $\mathbb{C}^2/m$, where m is a McKay group (a finite subgroup of $\mathcal{SU}(2)$). This desingularization is accomplished by blowing up the singlular point into a bouquet of 2-spheres, which has the contact structure determined by the Coxeter graph.

In the context of IIB string theory compactified on K3 spaces (which have an ALE structure), the gauge group for the vector multiplets is a product of unitary groups determined by the balance numbers of the corresponding extended Coxeter graph.

In the E_7 case, the balance numbers {1,2,3,4,3,2,1,2} determine the gauge group acting on the vector multiplets:

$$\mathcal{U}(1)^2 \times \mathcal{U}(2)^3 \times \mathcal{U}(3)^2 \times \mathcal{U}(4).$$

Moreover, the hypermultiplets living on the $\mathbb{C}^2/\mathcal{OD}$ space correspond exactly to the links on the extended E_7 Coxeter graph.

```
1—2—3—4—3—2—1
        |
        2
```

Thus we have the multiplets:

$$\mathbf{(1,2)} + \mathbf{(2,3)} + \mathbf{(3,4)} + \mathbf{(3,4)} + \mathbf{(2,3)} + \mathbf{(1,2)} + \mathbf{(2,4)}$$

so that the total dimensionality of this representation is:

$$2 + 6 + 12 + 12 + 6 + 2 + 8 = 48$$

which is the order of the $\mathcal{OD}$ McKay group.

Note that the links in this diagram correspond exactly to the quiver arrows in a quiver diagram cf. Chaps. 8 and 16.

The summary of ADE duality relations, invoked in Chap. 8, can usefully be repeated here:

1: McKay correspondence groups — $\mathcal{OD} \leftrightarrow$ affine E_7
2: ALE spaces (gravitational instantons) $\leftrightarrow \mathbb{C}^2/\mathcal{OD}$
3: Coxeter graph dual $\leftrightarrow$ bouquet of 2-spheres
4: McKay group algebra $\mathbb{C}[\mathcal{OD}] \leftrightarrow$ Matric basis $\leftrightarrow 8\ \mathcal{U}(n)$s
5: McKay group iireps $\leftrightarrow$ **IIB** hypermultiplets
6: Catastrophe germs $\leftrightarrow$ Superpotential $\leftrightarrow$ Higgsing SUSY
7: 2D conformal field theories $\leftrightarrow$ D-string worldsheet fields
8: K–M Lie algebras $\leftrightarrow$ Affine ADE Coxeter graphs
9: Finite Quivers $\leftrightarrow$ **IIB** hypermultiplets
10: Knots and links $\leftrightarrow AdS_5 \otimes S^5$

(9) In Chap. 9 we find the use of the extended E_7 Coxeter graph to describe the vertices of the $\mathcal{OD}$ polytope in $\mathfrak{R}^4$. These 48 vertices consist of two 24-cells reciprocal to each other. This $\mathcal{OD}$ polytope has 288 (non-regular) tetrahedral 3D faces. The dual $\mathcal{OD}$ polytope has 288 vertices and 48 truncated cube 3D faces [Coxeter, 1973; Duval, 1964].

These key numbers, 48 and 288 correspond to the extended E_7 Coxeter graph. The sum of squares of the balance numbers of this graph {1, 2, 3, 4, 3, 2, 1, 2} is:

$$1 + 4 + 9 + 16 + 9 + 4 + 1 + 4 = 48$$

and the product of the E_7 balance umbers is:

$$1 \times 2 \times 3 \times 4 \times 3 \times 2 \times 1 \times 2 = 288.$$

Note that a similar calculation works for the $\mathcal{TD}$ group with the Coxeter graph balance numbers, {1,2,3,2,1,2,1}. The 24 $\mathcal{TD}$ elements correspond to the 24 vertices of the 24 cell polytope in $\mathfrak{R}^4$. The 24 cells are octahedral faces, corresponding to the product of the $\mathcal{TD}$ balance numbers.

However, these are special cases. The sum of the ID balance numbers group yields the 120 vertices of the ID polytope. Yet the product of the ID balance numbers is 17,280, which does not match the 600 faces of the ID polytope [Coxeter, 1973].

It is important to view $\mathfrak{R}^4$ as the space of Hamilton's quaternion algebra $\mathcal{H}$, so that both $\mathcal{TD}$ and $\mathcal{OD}$ as finite subgroups of $\mathcal{SU}(2)$, and thus as McKay groups are here invoked.

Note also that $\mathcal{SU}(2)$, as a 3-sphere, is imbedded as the set of unit length quaternions in $\mathcal{H}$ cf. Chap. 4.

Another unifying theme in this book has been the emphasis on dualities, already invoked above. Here we will mention some of the more physically salient dualities. These dualities refer to two different mathematical descriptions which are equivalent in the sense that they provide the same physical (quantum mechanical) predictions.

The most important such dualities are those invoked by the M-theory of [Witten, 1995] and F-theory of [Vafa, 1996]. Here we can reproduce the iconic image of these dualities as described in Chaps. 8 and 15.

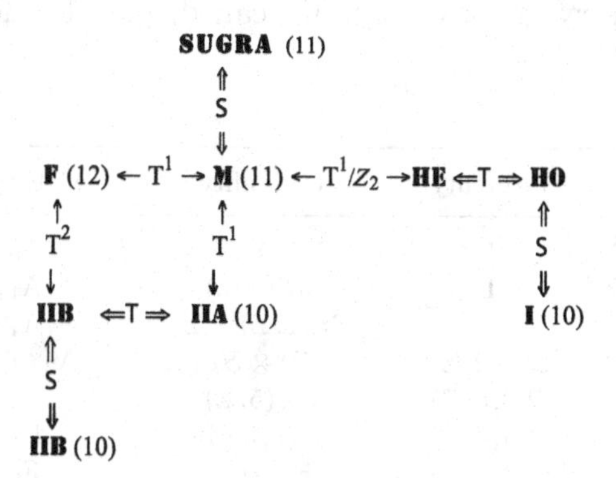

Here **I, IIA, IIB, HE** and **HO** are the five superstring theories unified as 10D subtheories of **M** (11D) theory which is a subtheory of **F** (12D) theory.

SUGRA is the 11D low-energy limit of M-theory.

$\mathcal{T}$ is T-duality, which exchanges size for string winding number on toroidal structures in compactification space.

$\mathcal{S}$ is S-duality, which exchanges weak and strong effects of the gauge coupling constants. Note that **IIB** has a self-dual form of S-duality.

T^1 is a 1D torus; T^2 is a 2D torus; T^1/Z_2 is a line-segment.

A more detailed description can be found in Chap. 8.

Moreover, there is also a *U-duality*, which is a combination of *T-duality* and *S-duality*. Thus *U-duality* is a symmetry between large toroidal radii with strong gauge coupling and small toroidal radii with weak gauge coupling.

In the case of **HE** and **HO**, the supergravity group corresponds to T-duality and U-duality groups of D-type running from D_8 through D_{16}. This is because these two heterotic theories are 26D theories, with bosonic strings active in 26D spacetime, while superstrings are active in 10D spacetime. The 16D Torus in the D_{16} type Lie group $\mathcal{SO}(32)$ interpolates between the two spacetimes, 26D and 10D [Kaku, 1999].

In the case of **IIA** and **IIB** Superstring theories, there is a series of (dimension dependent) symmetry groups of ADE type. There is an overall supergravity gauge group, as well as T-duality and U-duality symmetry groups. Thus we can display the following table from Chap. 14.

Dim.	T-duality	U-duality	ADE type
10A	$\mathbf{1}$	$\mathbf{1}$	A_0
10B	$\mathbf{1}$	$\mathcal{SL}(2,\mathbb{Z})$	A_1
9	$\mathbb{Z}_2$	$\mathcal{SL}(2,\mathbb{Z})\otimes\mathbb{Z}_2$	A_1
8	$\mathcal{O}(2,2,\mathbb{Z})$	$\mathcal{SL}(3,\mathbb{Z})\otimes\mathcal{SL}(2,\mathbb{Z})$	$A_2\oplus A_1$
7	$\mathcal{O}(3,3,\mathbb{Z})$	$\mathcal{SL}(5,\mathbb{Z})$	A_4
6	$\mathcal{O}(4,4,\mathbb{Z})$	$\mathcal{O}(5,5,\mathbb{Z})$	D_5
5	$\mathcal{O}(5,5,\mathbb{Z})$	$E_{6(6)}(\mathbb{Z})$	E_6
4	$\mathcal{O}(6,6,\mathbb{Z})$	$E_{7(7)}(\mathbb{Z})$	E_7
3	$\mathcal{O}(7,7,\mathbb{Z})$	$E_{8(8)}(\mathbb{Z})$	E_8
2	$\mathcal{O}(8,8,\mathbb{Z})$	$E_{9(9)}(\mathbb{Z})$	E_9
1	$\mathcal{O}(9,9,\mathbb{Z})$	$E_{10(10)}(\mathbb{Z})$	E_{10}

Here, 10A refers to the 10D type **IIA** theory, while 10B refers to **IIB** theory. This series of theories corresponds to various versions supergravity theory, which is a point-particle limit of 11D **M**-theory. The first column refers to the number of uncompactified dimensions, while the compactification dimension is indicated as the rank n of the Lie group, which corresponds to the torus T^n, the Cartan subgroup of the ADE type Lie group.

This series provides a context for the most important version of supergravity theory. That is 11D supergravity compactified on T^7, which is the Cartan subgroup of the E_7 Lie group. This leaves four uncompactified dimensions, which can be viewed as 4D Minkowski space.

Most importantly, this 11D supergravity theory is the low-energy limiting subtheory of the 11D M-theory compactified on T^7.

The supergravity gauge group for this theory is the non-compact Lie group $E_{7(7)}$. This 133D group has 63 compact dimensions, and 70 non-compact dimensions. This theory has the maximum number of supergravity generators, and is thus called $N = 8$ supergravity.

This $E_{7(7)}$ supergravity theory has the following particle content [Hull, 1985; de Wit and Nicolai, 1986].

1 graviton (spin 2)	— $\mathcal{SU}(8)$ singlet
8 gravitini (spin 3/2)	— fundamental $\mathcal{SU}(8)$ representation
28 vector fields (spin 1)	— adjoint rep. of $\mathcal{SO}(8) \subset \mathcal{SU}(8)$
56 fermions (spin $\frac{1}{2}$)	— fundamental $E_{7(7)}$ representation
70 scalars (spin 0)	— coset space $E_{7(7)}/\mathcal{SU}(8)$

cf. Chap. 14.

Outlook: Review of the conjectures in this book

The emphasis on E_7 structures has been followed throughout this book because of the special relationships between E_7 and the $\mathcal{OD}$ McKay group, and its factor group $\mathcal{O}$, which is the symmetric-permutation group S_4. These groups were the starting ideas of this book, see Chaps. 2 and 3.

(1) The primary conjecture of this book is that the structure of the S_4 indicates the existence of three families of fermions, with two quarks and two leptons in each family. These 12 particle types correspond to the 12 odd permutations of four things (as elements of S_4).

Likewise the 12 even permutations (i.e. the tetrahedral subgroup as the alternating-4 group) correspond to classes whose cardinalities

{1, 3, 8} provide a proto-structure for 1 photon, 3 weakons, and 8 gluons, which are Standard Model gauge particle types.

(2) My secondary conjecture is that the octahedral double group, $\mathcal{OD}$, corresponds to a template for the supersymmetry partners of the particles corresponding to the octahedral group, $\mathcal{O}$, which is the factor group $\mathcal{OD}/\pm 1$, which is $\mathcal{OD}/\mathbb{Z}_2$.

As evidence for this conjecture, we can mention that the McKay correspondence links $\mathcal{OD}$ to E_7 and that $E_{7(7)}$ contains the symmetry structures for 11D supergravity (as described above).

Presumably, the 3-family $\mathcal{OD}$ template of particles would be a lower energy symmetry-broken form of the $E_{7(7)}N = 8$ supergravity structures.

For example, the 56 spin-$\frac{1}{2}$ fermions are modeled by the 56-vertex polytope corresponding to the 56 E_7 weight vectors in the 7D reflection space. These 56 vertices project down to the 54-vertex polytope in the 6D reflection space of E_6. The $54 = 28 + \underline{28}$, which is viewed as a chiral represention.

Accordingly, we have the (by now) standard projections:

$$E_6(28 + \underline{28}) \Rightarrow D_5(16 + \underline{16}) \Rightarrow A_4(5 + \underline{5} \quad \text{and} \quad 10 + \underline{10}),$$

where A_4 is the $\mathcal{SU}(5)$ grand unified theory of [Georgi and Glashow, 1974].

Note that these fermion multiplets account for only 1 family of leptons and quarks (in three colors), and accounting for matter and anti-matter states.

Thus the three-family template, as provided by the octahedral group structure, would be complementary to this series of particle assignments cf. Chap. 8.

(3) Chap. 12 provides the context for the conjecture:

Since $744 = 3(248)$, the initial j-invariant 744 is quite significant. This is counter to the usual deletion of 744 from the standard j-invariant by writing:

$$j(\tau) - 744 = J(\tau) = q^{-1} + 196884q + 21493760q^2 \ldots.$$

The importance of 3(248) lies in the fact that 248 is the dimensionality of the E_8 Lie algebra. Also, since the E_8 lattice is self-dual, both the fundamental and adjoint representations of the E_8 Lie algebra are 248-dimensional.

Moreover, three copies of the E_8 lattice form a 24D lattice, which corresponds to the 24D-lattice of the Golay-24 code cf. Chap. 13.

(4) Also we find in Chap. 13 the fact that there are two quite well-known Gleason's Theorems.

Andrew Gleason (1921–2008) is more famous among physicists for "Gleason's Theorem" (1957), which was a precursor to John Bell's theorem (1964) on quantum entanglement. Among coding theorists, "Gleason's Theorem" (1970) classifies possible even self-dual error-correcting codes (which correspond to even self-dual lattices) cf. [Gleason, 1957, 1971; Bell, 1987; Bolker, 2009].

It is significant that the quantum entanglement of qubits plays a fundamental role in quantum information theory. Accordingly, quantum error correction codes will be necessary to ensure computational reliability. Thus we conjecture that there must be some intimate relationship between the two quite different "Gleason's Theorems."

Such a relationship is likely to be quite useful.

(5) In Chap. 16, we find the conjecture that there must be an ADE classification of mirror-pairs of C-Y spaces.

This is based on the description of C-Y three-fold as toroidal fibrations of a basic S^3. In this picture singular versions of T^3 are located on a network of points on S^3 [Yau and Nadis, 2010].

We are conjecturing that, since S^3 is the manifold underlying the $\mathcal{SU}(2)$ group, the McKay groups as finite subgroups of $\mathcal{SU}(2)$ would provide a classification of these network points locating singular toroidal structures. This, of course, would be the ADE classification of the McKay correspondence.

(6) Also in Chap. 16, we find the conjecture:

Since $S_4 \subset \mathrm{PSL}_2(7)$, the E_7 quartic (of genus 3) corresponds to the three-family structure of S_4.

Here, $\mathrm{PSL}_2(7)$ is the automorphism group of the genus 3 curve (over $\mathbb{C}$), which is called the Klein quartic curve. Also, $\mathrm{PSL}_2(7)$ is the simple group of order 168, which corresponds to the 168 shaded regions of Klein's quartic curve [Gray, 1982; Elkies, 1998]. There are also 168 unshaded regions, which corresponds to the fact that the group $\mathcal{SL}_{2(7)}$ of order 336 has $\mathrm{PSL}_2(7)$ as a factor group. However, S_4 is a maximal subgroup of $\mathrm{PSL}_2(7)$. Note that $168 = 7(24)$. Thus we have:

$$\begin{array}{ccccc} \mathcal{SL}_{2(7)}/\mathbb{Z}_2 = & \mathrm{PSL}_2(7): & |336| & \Rightarrow & |168| \\ \downarrow & \downarrow & \downarrow & & \downarrow \\ \mathcal{OD}/\mathbb{Z}_2 = & \mathcal{O}: & |48| & \Rightarrow & |24| \end{array}$$

cf. [Coxeter, 1991; Scholl *et al.*, 2002].

Moreover, the E_7 quartic, $x^3 + xy^3 = 0$, is a genus-3 manifold with 28 bitangents, corresponding to the 56D fundamental representation of the E_7 Lie algebra [Clemens, 1980; Hartshorne, 1977] cf. Chaps. 2, 3, 5, 7 and 14.

Coda

I first learned about H.S.M. Coxeter's graphs in 1974 from his book *Regular Polytopes*, 3rd Ed. (Dover, 1973). In 1977 I learned about Dynkin diagrams during a lecture at Stanford about the "Buildings" of Jacques Tits.

When I read the book by Tits, *Buildings of the Spherical Type and Finite B-N Pairs* (Springer-Verlag, 1974), I found out that the ADE classification of these newly discovered mathematical objects was central to this topic. The phrase "Spherical Type" refers to Coxeter's terminology for finite reflection groups. And indeed it was Tits who introduced the terms Coxeter group, Coxeter graph, and Coxeter number.

In this book, we have mentioned Jacques Tits on p. 137 in connection with the prime factors of the Monster group.

The intertwining of ADE structures can be further illustrated by descriptions of Toda lattices (as in M. Toda, *Theory of Nonlinear Lattices*, Springer-Verlag, 1981). This deals with non-linear forces acting on particles, such as those studied by Fermi, Pasta, and Ulam.

I will here make the (rather modest) conjecture that there are many other mathematical objects which can be classified by way of the ADE Coxeter graphs.

Two lines of poetry from Robert Browning have been my companion during the writing of this book:

"Ah, but a man's reach should exceed his grasp,
Or what's a heaven for?"

Bibliography

Aharony, O., Bergman, O., Jafferis, D. L., and Maldacena, J., $N = 6$ superconformal Chern–Simons–matter theories, M2-branes and their gravity duals, arXiv:0806.1218v4, 2008.
Akivis, M.A. and Rosenfeld, B.A., *Elie Cartan (1869–1951),* Amer. Math. Soc., 1991, p. 60.
Allcock, Daniel, Braid pictures for Artin groups, *Trans. Amer. Math. Soc.* **354:9**, 3455–3474, 2002.
Andrews, G.E., *The Theory of Partitions*, Addison-Wesley, 1976.
Anselmi, D., Billo, M., Fre, P., Girardello, L. and Zaffaroni, A., ALE manifolds and conformal field theories, *Int. J. Mod. Phys. A* **9**, 2007–3058, 1994.
Anglin, W.S., The square dynamic problems, *Amer. Math. Monthly* **97**, 120–124, 1990.
Arnold, V.I., *Singularity Theory,* Cambridge Univ. Press, 1981.
Arnold, V.I., Gusein-Zade, S.M. and Varchenko, A.N., *Singularities of Differentiable Maps, Vol. I,* Birkhäuser, 1985.
Arnold, V.I., *Catastrophe Theory,* Springer, 1986.
Aspinwall, P.S., K3 surfaces and string duality, arXiv:hep-th/9611137v5, 1999.
Atiyah, M.F., Real and complex geometry in four dimensions, in *The Chern Symposium 1979,* Springer-Verlag, 1980.
Atiyah, M.F., *The Geometry and Physics of Knots,* Cambridge Univ. Press, 1990.
Baez, J.C., The Octonions, *Bull. Amer. Math. Soc.* **39**, 145–205, 2002.
Barth, W., Peters, C. and Van de Ven, A., *Compact Complex Surfaces,* Springer-Verlag, 1984.
Becker, K., Becker, M. and Schwarz, J.H., *String Theory and M-Theory,* Cambridge Univ. Press, 2007.
Beiler, A.H., *Recreations in the Theory of Numbers,* 2nd Ed., Dover, 1966.
Bell, J.S., *Speakable and Unspeakable in Quantum Mechanics,* Cambridge Univ. Press, 1987.
Berlekamp, E.R., *Algebraic Coding Theory,* McGraw-Hill, 1968.
Bielawski, R., Twistor quotients of Hyperkähler manifolds, in *Quaternionic Structures in Mathematics and Physics*, World Scientific, 2001.
Binetruy, Pierre, *Supersymmetry: Theory, Experiment, Cosmology,* Oxford Univ. Press, 2006.

Birkhoff, G. and MacLane, S., *A Survey of Modern Algebra,* 3rd Ed., Macmillan, 1968.

Blau, S.K., A string-theory calculation of viscosity could have surprising applications, *Phys. Today,* May 2005.

Bleecker, D., *Gauge Theory and Variational Principles,* Addison- Wesley, 1981.

Bolker, E.D., Andrew M. Gleason 1921–2008, *Notices AMS,* **56:10**, 1236–1267, Nov. 2009.

Born, M., *The Born–Einstein Letters,* MacMillan, 1971.

Borcherds, R., Monstrous moonshine and monstrous Lie superalgebras, *Invent. Math.* **109**, 405–444, 1992.

Borcherds, R., What is the Monster?, *Notices of the AMS* **49:9**, 2002.

Borsten, L., Dahanayake, D., Duff, M.J., Ebrahim, H., and Rubens, W., Wrapped branes as qubits, *Phys. Rev. Lett.* **100**, 251602, 2008.

Brav, C. and Thomas, H., Braid groups and Kleinian singularities, *Math. Ann.* **351**:**4**, 1005–1017, 2011.

Brieskorn, E.V., Singular elements of semi-simple algebraic groups, in *Proc. Intern. Congress Math. Nice, Vol. 2,* pp. 278–284, Gauthier-Villars, 1970.

Brink, L. and Nielsen, H.B., A simple physical interpretation of the critical dimension of spacetime in dual models, *Phys. Lett.* **B45**, 332–336, 1973.

Bruce, J.W. and Giblin, P.J., *Curves and Singularities,* Cambridge Univ. Press, 1984.

Calabi, E., The space of Kähler metrics, *Proc. Int. Congr. Math. Amsterdam* **2**, 206–207, 1954.

Calabi, E., On Kähler manifods with vanishing canonical class, in *Algebraic Geometry and Topology,* Princeton Univ. Press, 1955.

Cappelli, A. and Zuber, J.-B., A-D-E classification of conformal field theories, arXiv:0911.3242.

Carmeli, Moshe, *Classical Fields: General Relativity and Gauge Theory,* Wiley, 1982.

Carroll, Sean, *The Particle at the End of the Universe,* Dutton, 2012.

Cheng, M.C.N. and Harrison, S., Umbral moonshine and K3 surfaces, arXiv:1406.0619v3, 2015.

Chisholm, C.D.H., *Group Theoretical Techniques in Quantum Chemistry,* Academic Press, 1976.

Clemens, C.H., *A Scrapbook of Complex Curve Theory,* Plenum, 1980.

Conway, J.H. and Norton, S.P., Monstrous moonshine, *Bull. Lond. Math. Soc.* **11**, 308–339, 1979.

Conway, J.H., Monsters and moonshine, *Math. Intelligencer* **2:4**, 165–172, 1980.

Conway, J.H., Curtis, R.T., S.P. Norton, Parker, R.A., Wilson, R.A., *Atlas of Finite Groups,* Oxford Univ. Press, 1985. Online at: http://www.mat.bham.ac.uk/atlas/v2.0/.

Conway, J.H., and Sloane, N.J.A., *Sphere Packings, Lattices and Groups,* Springer-Verlag, 1988.

Coxeter, H.S.M., Discrete groups generated by reflections, *Ann. Math.* **35**, 588–621, 1934.

Coxeter, H.S.M. and Moser, W.O.J., *Generators and Relations for Discrete Groups,* 2nd Ed., Springer-Verlag, 1965.

Coxeter, H.S.M., *Introduction to Geometry,* 2nd Ed., Wiley, 1969.

Coxeter, H.S.M., *Regular Polytopes,* 3rd. Ed., Dover, 1973.

Coxeter, H.S.M., *Regular Complex Polytopes,* 2nd Ed., Cambridge Univ. Press, 1991.

de Wit, B. and Nicolai, H., Local SU(8) invariance in d = 11 supergravity, in *Supersymmetry and its Applications,* Eds. G.W. Gibbons, S.W. Hawking and P.K. Townsend, Cambridge Univ. Press, 1986.

Dirac, P.A.M., The fundamental equations of quantum mechanics, *Proc. Roy. Soc.* A **109**, 642–653, 1926.

Dirac, P.A.M., *The Principles of Quantum Mechanics,* 4th Ed., Oxford Univ. Press, 1958.

Dixon, G.M., *Division Algebras,* Kluwer, 1994.

Douglas, M.R., and Moore, G., D-branes, quivers, and ALE instantons, arXiv:hep-th/9603167v1.

Duff, M.J. and Ferrara, S., E7 and the tripartite entanglement of seven qubits, *Phys. Rev.* **D 76**, 025018, 2007.

Duff, M.J., String triality, black hole entropy and Cayley's hyperdeterminant, *Phys. Rev.* **D 76**, 025017, 2007.

Duff, M., Black holes and qubits, *CERN Courier,* May 5, 2010.

Dunajski, M. and Mason, L.J., Twistor theory of hyper-Kähler metrics with hidden symmetries, arXiv:math/0301171v2, 2003.

Duncan, J.F.R., Griffin, M.J. and Ono, K., arXiv:1503.01472v1, 2015.

Durfee, A.H., Knot invariants of singularieties, in *Algegraic Geometry, Arcata 1974,* Amer. Math. Soc., 1975.

Dur, W., Vidal, G. and Cirac, J.I., Three qubits can be entangled in two inequivalent ways, *Phys. Rev.* **A 62**, 062314, 2000.

Duval, Patrick, *Homographies, Quaternions and Rotations,* Oxford Univ. Press, 1964.

Eddington, A.S., *The Mathematical Theory of Relativity,* 2nd Ed., Cambridge Univ. Press, 1924.

Eddington, A.S., *The Philosophy of Physical Science*, Cambridge Univ. Press, 1939.

Eguchi, T., Ooguri, H. and Tachikawa, Y., Notes on the K3 surface and the Mathieu group M24, arXiv:1004.0956v2, 2010.

Elkies, N., The Klein quartic in number theory, in *The Eightfold Way,* MSRI Publications Vol. 58, 1998.

Fischler, W. and Rajaraman, A., M(atrix) string theory on K3, arXiv:hep-th/9704123v1, 1997.

Flurry, Jr., R.L., *Symmetry Groups: Theory and Chemical Applications,* Prentice-Hall, 1980.

Freedman, D.Z., van Nieuwenhuizen, P. and Ferrara, S., *Phys. Rev. D* **13**, 3214 (1976); and *D* **14**, 912 (1976).

Frenkel, I.B., Lepowsky, J. and Meurman, A., A natural representation of the Fischer–Griess monster with the modular function J as character, *Proc. Nat. Acaad. Sci. USA.* **81**, 3256–3260, 1984.

Frenkel, I.B., Lepowsk, J., and Meurman, A., A moonshine module for the monster, in *Vertex Operators in Mathematics and Physics,* Eds. J. Lepowsky, S. Mandelstam, and I.M. Singer, Springer-Verlag, 1985.

Gannon, Terry, The Cappelli–Itzykson–Zuber A-D-E classification, *Rev. Math. Phys.* **12**, 739–748, 2000.

Gannon, Terry, Monsterous moonshine: The first twenty-five years, *Bull. Lond. Math Soc.* **38**, 1–33, 2006.

Gannon, Terry, Much ado about Mathieu, arXiv:1211.5531v2, 2013.

Georgi, H. and Glashow, S.L., *Phys. Rev. Lett.* **32**, 438 (1974).

Georgi, H., *Lie Algebras in Particle Physics,* Benjamin/Cummings, 1982.

Gibbons, G., Euclidean quantum gravity, in *The Future of Theoretical Physics and Cosmology,* Eds. G.W. Gibbons, E.P.S. Shellard and S.J. Rankin, Cambridge Univ. Press, 2003.

Gilmore, R., *Lie Groups, Lie Algebras and Some of Their Applications,* Wiley, 1974.

Gilmore, R., *Catastrophe Theory for Scientists and Engineers,* Wiley, 1981.

Gimon, E.G. and Johnson, C.V., K3 Orientfolds, *Nucl. Phys. B* **477**, 715, 1996.

Gleason, A.M., Measures on the closed subspaces of a Hilbert space, *J. Math. Mech.* **6**, 885–893, 1957.

Gleason, A.M., Weight polynomials of self-dual codes and the MacWilliams identities, in *1970 Act. Congr. Znt. Math.*, Vol. 3, pp. 211–215, Gauthier-Villars, 1971.

Goddard, P. and Thorn, C.B., Compatibility of the dual Pomeron with unitarity and the absence of ghosts in the dual resonance model, *Phys. Lett. B* **40**, 235–238, 1972.

Golay, M.J.E., Notes on digital coding, *Proc. I.R.E.* **37**, 657, 1949.

Golay, M.J.E., Binary Coding, *I.R.E. Trans. Inform. Theory* **PGIT-4**, 23–28, 1954.

Golubitsky, M. and Guillemin, V., *Stable Mappings and Their Singularities,* Springer-Verlag, 1973.

Golubitsky, M. and Schaeffer, D., *Singularities and Groups in Bifurcation Theory, Vol. I,* Springer-Verlag, 1985.

Gorenstein, Daniel, *Finite Simple Groups*, Plenum, 1982.

Gray, J., From the history of a simple group, *Math. Intelligencer* **4:2**, 59–67, 1982.

Gray, O., On the complete classification of unitary $N = 2$ minimal superconformal field theories, *Commun. Math. Phys.* **312:3**, 611–615, 2012.

Green, M. and Gross, D., Eds., *Unified String Theories,* World Scientific, 1986.

Green, M.B. and Schwarz, J.H., Anomaly cancellations in supersymmetric $D = 10$ gauge theory and superstring theory, *Phys. Lett.* **149B**, 117–122, 1984.

Green, M.B., Schwarz, J.H. and Witten, E., *Superstring Theory,* Vols. I & II, Cambridge Univ. Press, 1987.

Greene, B.R. and Plesser, M.R., Duality in Calabi–Yau moduli space, *Nucl. Phys. B* **338**, 15, 1990.

Griess, R., A construction of F1 as automorphisms of a 196,883-dimensional algebra, *PNAS* **78**, 689–691, 1981.

Griess, R., The friendly giant, *Invent. Math,* **69**, 1–102, 1982.

Gross, D.J. and Wilczek, F., *Phys. Rev. Lett.* **30**, 1343, 1973.

Gross, D.J., Harvey, J.A., Martinec, E. and Rohm, R., Heterotic string theory (I). The free heterotic string, *Nucl. Phys. B* **256**, 53–284, 1985.

Gross, D.J., On the uniqueness of physical theories, in *A Passion for Physics,* Eds. C. DeTar, J. Finkelstein, and C.-I. Tan, World Scientific, 1985.

Gross, D.J., in *Nobel Lectures: Physics 2001–2005*, Ed. G. Ekspong, World Scientific, 2008.

Grove, L.C. and Benson, C.T., *Finite Reflection Groups,* 2nd Ed., Springer-Verlag, 1985.

Guillemin, V. and Sternberg, S., *Symplectic Techniques in Physics,* Cambridge Univ. Press, 1984.

Hamming, R.W., *Coding and Information Theory,* Prentice-Hall, 1980.

Hankins, T.L., *Sir William Rowan Hamilton,* Johns Hopkins Univ. Press, 1980.

Hartle, J.B. and Hawking, S.W., Wave function of the universe, *Phys. Rev. D* **28**, 2960, 1983.

Hartshorne, R., *Algebraic Geometry,* Springer-Verlag, 1977.

Hawking, S.W., Black hole explosions, *Nature* **248**, 30–31, 1974.

Hawking, S.W., Breakdown of predictability in gravitational collapse, *Phys. Rev. D* **14**, 2460–2473, 1976.

Hawking, S.W., Gravitational Instantons, *Phys. Lett. A* **60**, 81, 1977.

Hawking, S.W., *A Brief History of Time,* Bantam, 1988.

Hawking, S.W., *The Universe in a Nutshell,* Bantam, 2001.

Heckman, J.J., Particle physics implications of F-theory, arXiv:1001.0577v2, 2010.

Heisenberg, W., *The Physical Principles of the Quantum Theory,* Univ. of Chicago Press, 1930.

Heisenberg, W., *Physics and Philosophy,* Harper, 1958.

Henkel, M., *Conformal Invariance and Critical Phenomena,* Springer-Verlag, 1999.

Hiller, H., *Geometry of Coxeter Groups,* Pitman, 1982.

Hori, K., Katz, S., Klemm, A., Pandharipande, R., Thomas, R., Vafa, C., Vakil, R., and Zaslow, E., *Mirror Symmetry,* Amer. Math. Soc., 2003.

Howe, Paul, The $N = 8$ supergravity on the mass shell, in *Supergravity,* Eds. P. van Nieuwenhuizen and D.Z. Freedman, North-Holland, 1979.

Hughston, L.P. and Ward, R.S., *Advances in Twistor Theory,* Pitman, 1979.

Hull, C.M., The construction of new gauged $N = 8$ supergravities, in *Supersymmetry in Physics,* Eds. V.A. Kostelecky and D.K. Campbell, North-Holland, 1985.

Hull, C.M. and Townsend, P.K., Unity of superstring dualities, *Nucl. Phys. B* **438**, 109–137, 1995.

Humphreys, J., *Introduction to Lie Algebras and Representation Theory,* Springer-Verlag, 1972.

James, G. and Kerber, A., *The Representation Theory of the Symmetric Group*, Addison-Wesley, 1981.

Jammer, M., *The Philosophy of Quantum Mechanics*, Wiley, 1974.

Johnson, C.V. and Myers, R.C., Aspects of type IIB theory on asymptotically locally Euclidean spaces, *Phys. Rev. D* **55:10**, 6382–6393, 1997.

Johnson, C.V., *D-branes,* Cambridge Univ. Press, 2003.

Johnson, C.V. and Steinberg, P., What black holes teach about strongly coupled particles, *Phys. Today,* May 2010.

Kac, V.G., Infinite-dimensional algebras. Dedekind's n-function, classical Möbius function and the very strange formula, *Adv. Math.* **30**, 85–136, 1978.

Kac, V.G., An elucidation of "Infinite dimensional...and the very strange formula" E8(1) and the cube root of the modular invariant, *J. Adv. Math.* **35**, 264–273, 1980.

Kac. V.G., *Infinite Dimensional Lie Algebras,* 2nd Ed., Cambridge Univ. Press, 1985.

Kaku, M., *Quantum Field Theory,* Oxford Univ. Press, 1993.

Kaku, M., *Introduction to Superstrings and M-theory,* 2nd Ed., Springer-Verlag, 1999.

Kallosh, R. and Linde, A., *Phys. Rev. D* **73**, 104033, 2006.

Kane, G., *Supersymmetry,* Perseus, 2000.

Kauffman, L.H., *On Knots,* Princeton Univ. Press, 1987.

Kauffman, L.H., *Knots and Physics,* World Scientific, 1991.

Kendig, K., *Elementary Algebraic Geometry,* Springer-Verlag, 1977.

Klarreich, E., Mathematicians chase moonshine's shadow, https://www.quantamagazine.org/20150312, 2015.

Klein, F., *Lectures on the Icosahedron,* Dover, 1956.

Klein, F., On the order-seven transformation of elliptic functions (translated by Silvio Levy), in *The Eightfold Way,* MSRI publications, Vol. 35, 1998.

Kokkedee, J.J.J., *The Quark Model,* Benjamin, 1969.

Kostant, B., On finite subgroups of SU(2), simple Lie algebras and the McKay correspondence, *Proc. Natl. Acad. Sci. USA* **81**, 5275–5277, 1984.

Kostant, B., The McKay correspondence, the Coxeter element and representation theory, *Asterisque, hors serie,* pp. 209–255, 1985.

Kostant, B., The Coxeter element and the branching law for the finite subgroups of SU(2), arXiv:math/0411142v1, 2004.

Kovtun, P., Son, D.T., and Starinets, A.O., *Phys. Rev. Lett.* **94**, 111601, 2005.

Kronheimer, P.B., A Torelli-type theorem for gravitational instantons, *J. Diff. Geom.* **29**, 685–697, 1989.

Kronheimer, P.B., The construction of ALE spaces as hyper-Kähler quotients, *J. Diff. Geom.* **29**, 665–683, 1989.

Kronheimer, P.B., A hyper-Kahlerian structure on coadjoint orbits of a semisimple complex group, *J. London Math. Soc.* (2) **42**, 193–208, 1990.

Kuzenko, S. M., Nonlinear self-duality in $N = 2$ supergravity, *JHEP* **06**, 012, 2012.

Lepowsky, J., Euclidean Lie algebras and the modular function, *J. Amer. Math. Soc. Proc. Symp. Pure Math.* **37**, 567–570, 1980.

Lepowsky, J. and Wilson, R.L., L-algebras and the Rogers–Ramanujan identities, in *Vertex Operators in Mathematics and Physics,* Eds. J. Lepowsky, S. Mandelstam and I.M. Singer, Springer-Verlag, 1985.

Levay, P., Strings, black holes, the tripartite entanglement of seven qubits and the Fano plane, *Phys. Rev. D* **75**, 024024, 2007.

Lincoln, D., The inner life of quarks, *Sci. Amer.* **22:2**, Summer 2013.

Lindstrom, U. and Rocek, M., Properties of hyperkahler manifolds and their twistor spaces, *Commun. Math. Phys.* **293**, 257–278, 2010.

Looijenga, E.J.N., *Isolated Singular Points on Complete Intersections,* Cambridge Univ. Press, 1984.

MacDonald, I.G., Affine root systems and Dedekind's eta-function, *Inventions Math.* **15**, 91–141, 1972.

MacWilliams, F.J. and Sloane, N.J.A., *The Theory of Error-correcting Codes,* North-Holland, 1977.

Maldacena, J., The large N limit of superconformal field theories and supergravity, *Adv. Theor. Math. Phys.* **2**, 231–252, 1998.

Mandelbrot, B.B., *The Fractal Geometry of Nature,* Freeman, 1983.

Martin, J.L., *Basic Quantum Mechanics,* Oxford Univ. Press, 1981.

Matsen, F.A., Frobenius algebras and the symmetric group, In *Group Theory and its Applications Vol. III,* Ed. E.M. Loebl, Academic Press, 1975.

McEliece, R.J., The reliability of computer memories, *Sci. Amer.*, Jan. 1985.

McKay, J., Graph, singularities, and finite groups, in *Proc. Symp. Pure Mathematics,* Vol. 37, pp. 183–186, 1980. Amer. Math. Soc., 1980.

McKay, J. and Ford, D., Representations and Coxeter graphs, in *The Geometric Vein: The Coxeter Festschrift,* Eds. C. Davis, B. Grunbaum, and F.A. Sherk, Springer-Verlag, 1981.

Milnor, J., On the 3-dimensional Brieskorn manifolds $M(p,q,r)$, in *Knots, Groups, and 3-Manifolds,* Ed. L.P. Neuwirth, Princeton Univ. Press, 1975.

Milnor, J., *Singular Points of Complex Hypersurfaces,* Princeton Univ. Press, 1968.

Misner, C.W., Thorne, K.S. and Wheeler, J.A., *Gravitation,* Freeman, 1973.

Moser, L., Elementary surgery along a torus knot, *Pacific J. Math.* **38**, 737–745, 1971.

Nambu, Y., in Proc. *Intern. Conf. on Symmetries and Quark Models,* p. 269, Gordon and Breach, 1970.

Okuda, T. and Ookouchi, Y., Higgsing and superpotential deformations of ADE superconformal theories, *Nucl. Phys. B* **733**, 59–90, 2006.

Orlik, P., *Seifert Manifolds,* Springer-Verlag, 1972.

Page, D., Quantum cosmology, in *The Future of Theoretical Physics and Cosmology,* Eds. G.W. Gibbons, E.P.S. Shellard and S.J. Rankin, Cambridge Univ. Press, 2003.

Pauli, W., The connection between spin and statistics, *Phys. Rev.* **58**, 716–730, 1940.

Peng, X.-H., On construction of the (24,12,8) Golay codes, *Infor. Th. IEEE Trans.* **52:8**, 3669–3675, 2006.

Penrose, R., The geometry of the universe, in *Mathematics Today,* Ed. L.A. Steen, Springer-Verlag, 1978.

Penrose, R. and Rindler, W., *Spinors and Space-time,* Vols. I, II, Cambridge Univ. Press, 1984; 1986.

Penrose, R., *The Road to Reality,* Knopf, 2005.

Polchinski, J., *String Theory* Vols. I, II, Cambridge Univ. Press, 1998.

Policastro, G., Son, D.T. and Starinets, O., Shear viscosity of strongly coupled $N = Y$ supersymmetric Yang–Mills plasma, *Phys. Rev. Lett.* **87**, 081601, 2001.

Politzer, H.D., *Phys. Rev. Lett.* **30**, 1346, 1973.

Politzer, H.D., *Phys. Rep.* **14C**, 129, 1974.

Porteous, I.R., *Topological Geometry,* 2nd Ed., Cambridge Univ. Press, 1981.

Preparata, F. P. and Vuillemin, J., The cube-connected cycles: a versatile network for parallel computation, *Communi. ACM*, **24**, 300–309, 1981.

Queen, L., Modular functions arising from some finite groups, *MTAC* **37**, 547–580, 1981.

Rauch, H.E. and Lewittes, J., The Riemann surface of Klein with 168 automorphisms, in *Problems in Analysis,* Ed. by R. C. Gunning, Princeton Univ. Press, 1970.

Rolfsen, D., *Knots and Links,* Publish or Perishs, 1976.

Ronan, M., *Symmetry and the Monster,* Oxford Univ. Press, 2006.

Ross, G.G., *Grand Unified Theories,* Benjamin/Cummings, 1984.

Schiff, L.I., *Quantum Mechanics,* 3rd Ed., McGraw-Hill, 1968.

Scherk, J. and Schwarz, J.H., *Nucl. Phys. B* **81**, 118, 1974.

Scholl, P., Schurmann, A. and Wills, J.M., Polyhedral models of Felix Klein's group, *Math. Intelligencer* **24:3**, 37–42 2002.

Schutz, B., *Geometrical Methods of Mathematical Physics,* Cambridge Univ. Press, 1980.

Schwarz, J.H., From the bootstrap to superstrings, in *A Passion for Physics: Essays in Honor of Geoffrey Chew,* World Scientific, 1985.

Seidel, P. and Smith, I., A link invariant from the symplectic geometry of nilpotent slices, *Duke Math. J.* **134**, 453–514, 2006.

Seidel, P., Lagrangian two-spheres can be symplectically knotted, *J. Diff. Geom.* **52**, 145–171, 1999.

Shannon, C.E., The mathematical theory of communication, *Bell Syst. Tech. J.,* **27**, 379–423, 1948; **27**, 623–659, 1948.

Shannon, C.E., Communication in the presence of noise, *Proc. IRE* **37**, 10–21, 1949.

Shephard, G.C. and Todd, J.A., Finite unitary reflection groups, *Canad. J. Math.* **6**, 274–304, 1954.

Sirag, S.-P., Why there are three fermion families, *Bull. Amer. Phys. Soc.* **27**, 31, 1982.

Sirag, S.-P., A finite group algebra unification scheme, *Bull. Amer. Phys. Soc.,* **34**, 82, 1989.

Sirag, S.-P., Consciousnes: a hyperspace view, in *The Roots of Consciousness,* 2nd Ed., Jeffrey Mishlove, Marlowe, 1993.

Sirag, S.-P., A mathematical strategy for a theory of consciousness, in *Toward a Science of Consciousness,* Eds. S.R. Hameroff, A.W. Kaszniak, and A.C. Scott, pp. 580–588, MIT Press, 1996.

Sloane, N.J.A., Self-dual codes and lattices, *Proc. Symp. in Pure Math.* **34**, 1979.

Sloane, N.J.A., Error-correcting codes and cryptography, in *The Mathematical Gardner,* Ed. D.A. Klarner, Wadsworth International, 1981.

Sloane, N.J.A., Gleason's theorem on self-dual codes and its generalizations, arXiv:math/0612535v1, 2006.

Sloane, N.J.A., http://oeis.org/A106318

Slodowy, P., Platonic solids, Kleinian singularities, and Lie groups, in *Algebraic Geometry,* Ed. I. Dolgachev, Springer-Verlag, 1983.

Son, D., Liquid univers hints at strings, *Phys. World,* June 2005.

Steinberg, R., Finite reflection groups, *Amer. Math. Soc. Trans.* **91**, 493–504, 1959.

Stekolshchik, R., Kostant's generating functions, Ebeling's theorem and McKay's observation relating the Poincaré series, arXiv:math/0608500v1, 2006.

Stembridge, J., Coxeter planes, http://www.math.lsa.umich.edu/~jrs/coxplane.html, 2007.

Streater, R.F. and Wightman, A.S., *PCT, Spin and Statistics, and All That,* Benjamin, 1964.

Strominger, A., Yau, S.-T. and Zaslow, E., Mirror symmetry is T-duality, *Nucl. Phys. B* **479**, 243–259, 1996.

Susskind, L., *Nuovo Cimento* **69 A**, 457, 1970.

Susskind, L., The world as a hologram, *J. Math. Phys.* **36**, 6377–6396, 1995.

Susskind, L., Twenty years of debate with Stephen, in *The Future of Theoretical Physics and Cosmology,* Cambridge Univ. Press, 2003.

Susskind, L. and Lindesay, J, *Black Holes, Information and the String Theory Revolution,* World Scientific, 2005.

't Hooft, G., A planar diagram theory for strong interactions, *Nucl. Phys. B* **72**, 461, 1974.

't Hooft, G., Dimensional reduction in quantum gravity, in *Salamfestschrift,* p. 284, World Scientific, 1994 arXiv:gr-qc/9310026.

Thompson, D.W., *On Growth and Form,* Cambridge Univ. Press, 1942.

Thom, R., *Structural Stability and Morphogenesis*, Benjamin, 1975.

Tyurina, G.N., Resolutions of singularities of plane deformations of rational double points, *Funct. Anal. Appl.* **4**, 68–73, 1970.

Vafa, C., Evidence for F-theory, *Nucl. Phys. B* **469**, 403, 1996.

Van der Lek, H., Extended Artin groups, in *Proc. Symp. Pure Math.* **40**, Part 2, 1983.

van der Waerden, B.L., *Sources of Quantum Mechanics,* Dover, 1968.

van Nieuwenhuizen, P. and Freedman, D.Z., *Supergravity*, North-Holland, 1979.

van Nieuwenhuizen, P., Supergravity, in *Some Strangeness in the Proportion,* Addison-Wesley, 1980.

Veltman, M., *Facts and Mysteries in Elementary Particle Physics,* World Scientific, 2003.

Waddington, C.H., *The Strategy of the Genes: A Discussion of Some Aspects of Theoretical Biology,* Allen & Unwin, 1957.

Wald, R.M., *General Relativity,* Univ. of Chicago Press, 1984.

Ward, R.S. and Wells, Jr., R.O., *Twistor Geometry and Field Theory,* Cambridge Univ. Press, 1990.

Warner, N., Gauged supergravity and holographic field theory, in *The Future of Theoretical Physics and Cosmology,* Eds. G.W. Gibbons, E.P.S. Shellard and S.J. Rankin, Cambridge Univ. Press, 2003.

Waltron, G.N., The problem of the signal pyramid, *Messenger Math.* **48**, 1–22, 1918.

Weeks, J.R., *The Shape of Space,* Dekker, 1985.

Weinberg, S., *Gravitation and Cosmology,* Wiley, 1972.

Weinberg, S., *Cosmology,* Oxford Univ. Press, 2008.

Wess, J. and Bagger, J. *Supersymmetry and Supergravity,* Princeton Univ. Press, 1983.

Weyl, H., *The Theory of Groups and Quantum Mechanics,* Dover, 1950.

Witten, E., String theory dynamics in various dimensions, *Nucl. Phys. B* **443**, 85–126 1995.

Witten, E., Refletions on the fate of spacetime, *Phys. Today* **49**, 24–30, April 1996.

Witten, E., Anti de Sitter space and holography, *Theor. Math. Phys.* **2**, 1113–1133, 1998.

Yau, S.-T., Calabi's conjecture and some new results in algebraic geometry, *Proc. Natl. Acad. Sci. USA* **74**, 1798–1799, 1977.

Yau, S.-T. and Nadis, S., *The Shape of Inner Space,* Basic Books, 2010.

Zee, A., *Unity of Forces in the Universe,* Vols. I & II, World Scientific, 1982.

Zuber, J.-B., CFT, BCFT, ADE and all that, *Contemporary Math.* **294**, 233–266, 2002.

Zwiebach, B., *A First Course in String Theory,* Cambridge Univ. Press, 2004.

Glossary

Affine space. A flat space, whose properties are intermediate between Euclidean and Projective. Cf. [Coxeter, 1969].

Affine Dynkin diagram. The diagram for an infinite order Coxeter group, and also for an infinite dimensional Lie algebra. Such diagrams have numbers attached to the nodes. I have called these **balance numbers**. The infinite dimensional Lie algebras are called **Affine Kac–Moody** Lie algebras. Cf. Chap. 4 and [Kac, 1985].

ALE space. An asymptotically locally Euclidean space. These manifolds were ADE classified by Peter Kronheimer. They are equivalent to the set of **Gravitational instantons.**

There is a close relationship between ALE spaces and the K3 space, which is a Calabi–Yau 2-fold. Both are Kähler spaces with a Ricci flat metric. However, while an ALE space is non-compact, K3 is compact. Thus an ALE space is a decompactified form of K3 space, which has an ADE classified set of **orbifold** limits: $\mathbb{C}^2/m$, where m is a finite subgroup of $\mathcal{SU}(2)$, the holonomy group of K3. Cf. Chaps. 8 and 16.

Algebra. The study of the structure of numbers and the systems of entities abstracted from and generalized upon this structure. The most basic structure is the *group*, both additive (which is commutative) and multiplicative (which may be non-commutative). A **linear algebra** can be represented by matrices, which can be added and multiplied. Thus the addition and multiplication properties (and also distribution between addition and multiplication) of the ordinary algebra of real and complex numbers can be generalized by such a matrix algebra.

Moreover, any matrix algebra can be made into a **Lie algebra** by defining Lie brackets on the underlying matrix algebra:

$[a, b] = ab - ba$, where ab and ba are matrix multiplications. Cf. **Group algebra.**

Balance number. The numbers attached to the affine Dynkin diagram. Such an affine diagram has $n+1$ nodes for a Lie algebra of rank n. The extra node is placed on the graph so that in an n-dim. vector space, vectors attached to the origin, having lengths corresponding to the node numbers will sum to zero. Using the analogy of force vectors, the force would sum to zero, and thus the vectors would be in balance, suggesting the node numbers as **balance numbers.** The sum of the balance numbers is the **Coxeter number** K. Another way to see this is that (excluding the extra node), the numbers on the remaining n nodes of an affine diagram are the components of the **highest weight**, so that the sum of these components plus 1 is the Coxeter number K.

Basis. A set **B** of *vectors* in a vector space such that any vector in the space is equivalent to a sum over **B**. The cardinality of **B** is the dimensionality of the vector space.

Bell's theorem. An inequality, published by John Bell in 1964. This inequality is violated by the predictions of quantum mechanics. This inequality would hold if two particles in a *singlet state* could (when widely separate from each other) communicate faster than the speed of light. Experiments beginning in the '70s show that Bell's inequality is violated. Thus the predictions of quantum mechanics are upheld; and so this strange property called **entanglement** is a fundamental aspect of the physical world.

Brane. A generalization of the 2D membrane to describe surfaces of any dimension p. Thus Branes are called p-branes, so that in 10D spacetime p can range from 0 to 9. A string is thus a 1-brane.

An important special case is the **D-brane**, which can serve as a boundary for open strings. These strings are attached to the D-brane, or have both ends of the string attached to separate D-branes. D-branes come in various dimensions and are thus called D_p-branes. The D stands for Dirichlet, because D-branes obey Dirichlet boundary conditions.

Closed strings do not attach to D-branes and can move freely in the larger bulk space in which the D-branes reside. Closed strings correspond to **gravitons**.

Branes of various types are important in supertring theories of types **I, IIA** and **IIB**, as well as in $\mathcal{M}$-theory which unifies all five types of Superstring theory as 10D sub-theories of the 11D $\mathcal{M}$-theory.

Calabi–Yau space. A Kähler manifold with first Chern class equal to zero. It is called a CY n-fold, or CY(n), because it has n complex dimensions. Moreover the CY(n) has a Ricci-flat metric of $\mathcal{SU}(n)$ holonomy.

In superstring applications, the CY(3) and its **mirror symmetry** is the most studied case. This is because CY(3) is a 6D real space which can be the compactified space of the 10D spacetime of superstring theory.

The CY(2) is also applied in string theories compactified on **K3** spaces. There are two types of CY 2-folds: T^4 and K3. Moreover, K3 can be described by way of its various orbifold limits $\mathbb{C}^2/\mathfrak{m}$, where $\mathfrak{m}$ is a finite subgroup of $\mathcal{SU}(2)$, and is thus a McKay group. In this sense K3 spaces have ADE classified orbifold limits, since the McKay groups are ADE classified. Cf. **ALE spaces.**

Conformal field theory. A quantum field theory that obeys conformal invariance. Such a theory remains invariant under a conformal mapping which preserves direction but neglects to preserve distance.

In string theory the 2D worldsheet swept out by the strings is conformally invariant. Thus the topology of the worldsheet is all important.

Coxeter arrangement. The set of mirror hyperplanes generated by the basic mirrors corresponding to the nodes of a **Coxeter graph**. The number of such hyperplanes in such a Coxeter arrangement is $nK/2$, where n is the rank (the number of nodes) of the Coxeter graph, and K is the **Coxeter number**. The n basic mirrors generate the **Coxeter group**.

Coxeter element. For a **Coxeter group** generated by n **reflections**, the Coxeter element S is the product of all n reflections. Since this group is non-commutative, there is a family of such Coxeter elements. However, they are all conjugate to each other, and thus have the same **order**, which is the Coxeter number K. The order of any group element g is the smallest number k such that $g^k = e$, the identity element. Thus we can write for the Coxeter element: $s^K = e$, so that K is the Coxeter number. The sum of the balance numbers of the affine Dynkin diagram is also K.

Coxeter graph. A set of nodes connected by links, in which each node corresponds to a basic reflection plane, and each link corresponds to an angle π/n (where $n \geq 3$). The angle π/n holds between the basic mirror planes which generate the **Coxeter reflection group**. By convention, the links with $n = 3$ (corresponding to 60°) are left unmarked, because they are so common. The ADE graphs have only $n = 3$, and so are unmarked. For the full set of Coxeter graphs, see Chap. 6.

The ADE Coxeter graphs are also called **singly laced Dynkin diagrams**. This is because the Dynkin diagrams (which classify the simple Lie algebras) are closely related to the Coxeter graphs, but use a different marking convention for the links: multiple lines and arrows. Thus the singly laced Dynkin diagrams are of ADE type and correspond exactly to the unmarked Coxeter graphs.

Coxeter groups. Finite reflection groups, which are classified by the complete set of **Coxeter graphs.** The crystallographic reflection groups correspond to the **Weyl groups** of simple Lie algebras. There are finite reflection groups (and thus Coxeter groups) which are not crystallographic. Cf. Chap. 6.

Coxeter number. The number K which is the order of the **Coxeter element**. K is also the sum of the **balance numbers** on the **Affine Coxeter graph**.

Deformation. A transformation of a geometrical structure which shrinks, twists or otherwise changes the structure without tearing.

Dimension. The number of degrees of freedom in a space, which is the number of coordinates necessary to locate a point in a space. The coordinates are not necessarily rectangular.

Duality. A quantum physical equivalence between two different mathematical formalisms. The first such example was the *wave-particle duality*. Superstring theory affords many more such quantum dualities between various versions of string theory.

Mirror symmetry is a duality between pairs of Calabi–Yau 3-folds. Here the winding number of the string in one CY(3) is dual to the quantized momentum of the string in the dual CY(3), and vice versa.

T-duality. Mirror symmetry is an example of T-duality, because the string winds on a torus in the CY(3). Note that the T means toroidal. Among the various types of superstring theory, T-duality corresponds to toroidally compactified theories. Thus there is a T-duality between type **IIA** and **IIB** theories. Also a T-duality holds between **Het.E** and **Het.O** theories.

S-duality. This is a duality between weak and strong coupling constants, so that what is very difficult to calculate at strong coupling becomes easy to calculate at weak coupling. Notably, **IIB** has an S-duality between weak and strong versions of itself. So this is called a self S-duality. There is also an S-duality between the type **I** theory and the **Het.O** theory. Most notably, there is an S-duality between M-theory and the 11D **supergravity** theory, which is the low energy limit of M-theory. Cf. Chap. 18.

Dynkin diagram. This is a set of graph-like structures. Each diagram is a set of nodes connected with links which can be singly laced (in the case of the ADE diagrams); doubly or triply laced. The Lie algebras of type B and C correspond to diagrams with one doubly laced link. The F_4 diagram has one double link and two single links between its four nodes. The G_2 diagram has a triple link between its two nodes.

The singly laced Dynkin diagrams correspond exactly to the ADE Coxeter graphs. Note, however, that the **Coxeter graphs** do not distinguish between B and C type structures. This is because B and C type Lie algebras correspond to the same **Coxeter group** (also called the **Weyl group**). $\mathrm{B_n}$ type Lie algebras correspond to $\mathcal{SO}(2n+1)$ Lie groups; and $\mathrm{C_n}$ type Lie algebras correspond to $\mathrm{Spl}(n)$ Lie groups. Note also that the $\mathcal{SO}(2n)$ Lie groups are of $\mathrm{D_n}$ type.

Moreover, **Affine Coxeter graphs** do distinguish between $^{\infty}\mathrm{B_n}$ and $^{\infty}\mathrm{C_n}$ type graphs. Cf. [Coxeter and Moser, 1965].

Eigenvalue. The numerical solution x to the operator equation:

$$\mathcal{O}\mathbf{v} = \boldsymbol{x}\mathbf{v},$$

where $\mathcal{O}$ is the operator, and $\mathbf{v}$ is a vector called the *eigenvector* solution to this equation. Geometrically, this means that $\mathcal{O}$ is acting upon the vector $\mathbf{v}$ and does not change its direction but only its length; this length change is by the factor $\boldsymbol{x}$.

Usually $\mathcal{O}$ is represented by an $n \times n$ matrix, where n is the dimension of $\mathbf{v}$, i.e. the number of components in $\mathbf{v}$.

Eigenvector. A vector solution $\mathbf{v}$ to the operator equation described under *eigenvalue*. In general, there are several eigenvector solutions to an operator equation. The number n of solutions is the dimensionality of the space the vectors live in.

Entanglement. A quantum effect which correlates the properties of two or more particles, so that the measurement of one particle, affects state of the other particle (or particles). This effect is instantaneous and does not diminish over distance, and is thus called *nonlocality*. Einstein famously called it "spooky actions at a distance" [Born, 1971, p. 158]. However, this entanglement effect cannot be used to transfer information or energy faster than light speed. Cf. **Bell's theorem.**

Field (geometric). A smooth assignment of some type of mathematical object to each point of some space. For example, a *scalar* field assigns a scalar (a number, real or complex) to each point of a space. A *vector* field assigns a vector to each point of a space. A *tensor* field assigns a tensor (of rank r) to each point of a space. A vector is a rank 1 tensor. The tensors used in general relativity are rank 2, and are represented by 4×4 matrices (for 4D spacetime).

Field (algebraic). A structure **F** generalizing the algebra of real numbers. The elements of **F** obey two binary operations (called addition and multiplication). **F** has an additive identity element (called **zero**); and **F** has a multiplicative identity element (called **one**). **F** is a *Commutative Group* under addition. **F** (excluding zero) is a *Commutative Group* under multiplication. A *Distributive* relationship holds between multiplication and addition.

There are only two fields of infinite order: *Real* numbers, and *Complex* numbers. *Quaternions* are not a field, because they are not multiplicatively commutative.

There is, however, an infinity of *finite fields.* For each prime number p and positive integer n, there is a field of p^n elements. The multiplicative group is the cyclic group of order $p^n - 1$. The additive group of order p^n includes the zero element.

These finite fields are also called *Galois fields*, because Galois used the four-element field in proving that 5th degree polynomials have no general solution. Galois fields are denoted $GF(p^n)$.

Force field (also called a *gauge field).* A smooth assignment of an element of a Lie group to each point of spacetime. The Standard Model correspondence of Lie groups to force fields is as follows: $\mathcal{U}(1)$, electromagnetism; $\mathcal{SU}(2)$, weak force; $\mathcal{SU}(3)$, strong (color) force.

Function. A mapping from an n-dimensional space to a 1-dimensional space. A *linear function* is an additive entity in the sense that the sum of two linear functions on a space is also a linear function on that space.

Gravitational instanton. A solution to the Euclideanized vacuum field equations of general relativity. This is the desingularization of the 4D real space $\mathfrak{R}^4/\boldsymbol{m}$ whose boundary at infinity is $S^3/\boldsymbol{m}$ (rather than S^3). Here $\boldsymbol{m}$ is a finite subgroup of $\mathcal{SU}(2)$, and is thus a McKay group. The $S^3/\boldsymbol{m}$ boundary at infinity means that this space is asymptotically locally Euclidean, and is thus called an **ALE space**. The desingularization is accomplished by blowing up the singularity in $\Re^4/\mathfrak{m}$ into a bouquet of two-spheres corresponding to the ADE graph associated with the McKay group $\mathfrak{m}$. This gravitational instanton 4D space is very special. It has three complex structures defined on it. It is also a hyper-Kähler manifold, because it has three symplectic forms defined on it. The three complex structures are labeled I, J and K, because they obey Hamilton's Quaternion formula: $I^2 = J^2 = K^2 = IJK = -1$.

These gravitational instanton 4D spaces are decompactifications of the **K3**, Calabi–Yau 2-fold. This is because, like CY(2), they are Kähler manifolds with zero Ricci curvature (as vacuum solutions to 4D Euclideanized general relativity). Cf. **Singularities of ADE type.**

Graviton. A spin-2 particle of quantum gravity. In string theory, closed strings are spin-2 states, while open strings correspond to spin states 0, 1/2, 1 and 3/2. The spin-3/2 state corresponds to the **gravitino**, which is the fermionic supersymmetry partner to the bosonic graviton.

Group. A set G of elements obeying four laws:
Closure: for $\{a, b\} \in G, ab = c$ implies $c \in G$
Associativity: $a(bc) = (ab)c$
Identity element: $ea = ae = a$ implies $e =$ identiy
Inverse elements: $aa^{-1} = a^{-1}a = e$, for all $a \in G$
Commutativity: a 5th law obeyed by **commutative groups**

$$ab = ba.$$

Group algebra. Any finite group G of order $|G|$ can be made into the basis elements of a vector space (real or complex) of dimension $|G|$. This vector space has the additive property of its vectors; but it also has multiplicative properties induced by the underlying group G basis. Thus it is an associative (usually non-commutative) algebra called the group algebra of G. Over complex numbers, this group algebra is symbolized as $\mathbb{C}[G]$. The regular representation of dimension $|G|$ is isomorphic to a direct sum of total matrix algebras, which contain the matrix representations corresponding to the irreducible representations of the underlying finite group G. For example, in the case of the octahedral double group $\mathcal{OD}$, of order 48, we have:

$$\mathbb{C}[\mathcal{OD}] = M(1)+M(2)+M(3)+M(4)+M(3)+M(2)+M(1)+M(2)$$

and we note that $1^2+2^2+3^2+4^2+3^2+2^2+1^2+2^2 = 48$, so that the sum of the total matrix algebras spans the 48D vector space whose basis is the $\mathcal{OD}$ group of order 48.

Note that $\mathcal{OD}$ corresponds to the E_7 **affine Lie group**, whose Dynkin diagram has balance numbers corresponding to the irreducible representationos of the $\mathcal{OD}$. This correspondence is called the **McKay correspondence**, which is an ADE classification scheme. Thus all the finite subgroups of $\mathcal{SU}(2)$ have a **group algebra** structure analogous to that of $\mathcal{OD}$.

Heterotic string theory. A subtle weaving together of the 26D bosonic string theory and the 10D Superstring theory. The 16-dimensional difference between these two theories requires a gauge group of rank 16. Thus $E_8 \times E_8$ and D_{16} (corresponding to $\mathcal{SO}(32)$) are the gauge symmetry groups for the two versions of the heterotic theory. Note that both of these theories describe only closed strings, which are thus only gravity theories.

Hyperplane. A flat space of dimension $n - 1$ in a flat space of dimension n. For example, a 3D vector space as a subspace in a 4D vector space.

Hyperspace. A space of more than three dimensions. Physical examples include 4D Minkowski space, and the 10D spacetime of superstring theory.

Kähler manifold. Named for Erich Kähler, it is a manifold with n complex dimensions such that its $\mathcal{SU}(n)$ holonomy group preserves the complex structure during parallel transport of vectors around closed loops.

Lie algebra. A non-associative *algebra* which obeys two extra rules:

(1) Anticommutativity: $ab = -ba$
(2) Jacobi identity: $a(bc) = (ab)c + b(ac)$,
which replaces the associative law $a(bc) = (ab)c$.

Since most Lie algebras can be faithfully represented by sets of matrices, the Lie product can be faithfully represented by the

Lie bracket: $[a, b] = ab - ba$,

where the product ab is the ordinary product of two matrices.

Also, any matrix algebra is an associative algebra which can be made into a Lie algebra by defining Lie brackets on the underlying matrix algebra.

The building blocks of Lie algebras are *simple Lie algebras.*

M-theory. An 11D spacetime theory which contains the five versions of superstring theory as 10D limiting subtheories. M-theory does not itself contain strings, but rather 2D and 5D membranes. It is most closely related to **IIA** theory, which becomes the 11D M-theory by adding the torus T^1 to the 10D spacetime of the **IIA** theory.

Note also that the low energy limit of M-theory is the 11D supergravity theory, which is a point particle theory.

Mapping. A rule which assigns to each element of an n-dimensional space N a unique element of X (an x-dimensional space). Note that N and X may be the same space. A *function* is a special case of a mapping, in which X, the space being mapped to, is 1D.

Matter field. A smooth assignment of a complex vector (of appropriate dimension) to each point of spacetime. The vector space is acted upon by the Lie group associated with the *force field* involved.

McKay correspondence. The correspondence between finite subgroups of $\mathcal{SU}(2)$ and the ADE classified set of affine Lie algebras. The Dynkin diagrams of these affine Lie algebras have **balance numbers** on their nodes; and these balance numbers correspond to the irreducible representations of the corresponding finite subgroup of $\mathcal{SU}(2)$. For this reason I have labeled these subgroups $\mathcal{m}$ and called them **McKay groups.** Cf. Chap. 4.

Operator. A mathematical entity that transforms a space. For example, on a *vector space* an operator may rotate, stretch, translate or perform some combination of these transformations on the vectors of the space. An operator which acts on an n-dimensional space is represented by an $n \times n$ matrix.

In quantum mechanics, *Hermitian* operators play a fundamental role. This is because the *Wave function* is a complex function. The squares of the absolute value of the complex numbers are interpreted as the probabilities of observation. *Hermitian* operators are represented by complex matrices which have real eigenvalues. And it is these eigenvalues whose probabilities are predicted. Such *Hermitian* operators are called the quantum observables.

Orbifold. Given a manifold M acted on by a group G, the set of G-orbits in M is symbolized as M/G, and is called an orbifold of M.

Orbit. The set of points in a space selected by the action of all the elements of a group on a single point of the space. If the group is continuous (a Lie group), the orbit will be continuous. If the group is discrete, the orbit of the point will be a discrete set of points. For example, the orbit of a *reflection group* acting on a point between the mirrors (i.e. in a reflection chamber) is a point in each reflection chamber. Thus the number of points in such an orbit is equal to the number of elements in the reflection group.

Parameter. A variable which selects members of a family of structures. For example, a and b are parameters in the equation for a straight line: $y = a + bx$, so that a selects the point where the line intercepts the y-axis, and b selects the slope of the line.

In this book, in the case of the E_7 *catastrophe*: $t_1, \ldots, t_7$ are parameters which select a fiber of the E_7 catastrophe manifold X^9. These fibers are called the *deformations* of the identity fiber $C^2/\mathcal{OD}$. Cf. **Unfolding**.

Quaternion. An ordered pair of complex numbers in $\mathbb{C}^2$, as a generalization of complex numbers. Just as unit length complex numbers correspond to rotations in $\mathfrak{R}^2$, so unit length quaternions correspond to rotations in $\mathfrak{R}^4$. Quaternions are not commutative, but do have multiplicative inverses. Thus they are elements of a **division algebra**. A quaternion q has four components, such that:

$$q = q_0 1 + q_1 I + q_2 J + q_3 K,$$

where q_i are real numbers and $1, I, J, K$ are basis vectors in the 4D vector space of the quaternion algebra $\mathcal{H}$. These basis vectors are defined by Hamilton's formula:

$$I^2 = J^2 = K^2 = IJK = -1.$$

Moreover, the set of unit length vectors in $\mathcal{H}$ is the Lie group $\mathcal{SU}(2)$, whose geometric structure is the 3-sphere S^3. Also, the set of quaternions $\{\pm 1, \pm I, \pm J, \pm K\}$ constitutes an eight element non-commutative group called the quaternion group. This group is a subgroup of $\mathcal{SU}(2)$, and is thus the **McKay group** corresponding to the Coxeter graph labeled D_4 and to the Lie group $\mathcal{SO}(8)$.

For physicists, quaternions are closely related to **Pauli spin matrices**, which can be considered as the basis elements of the $\mathfrak{su}(2)$ Lie algebra.

Schrödinger equation. An eigenvalue equation for the **wave function.** In its simplest form it is:

$$(ih/2\pi)\partial\Psi/\partial t = H\Psi,$$

so that Ψ is the eigenfunction, with H (the Hamiltonian) representing the eigenvalue, which is the **energy**. Thus H is a quantum **observable.**

Singularities of ADE type. As classified by V.I. Arnold, these are simple singularities. They correspond to the ADE graphs in the sense that the singular variety $\mathbb{C}^2/\mathcal{m}$ consists of the orbits of $\mathcal{m}$, where $\mathcal{m}$ is a finite subgroup of $\mathcal{SU}(2)$, and is thus a **McKay group** (classified by the ADE graphs). Each ADE graph has a dual structure consisting a bouquet of two-spheres matching the nodes of the ADE graph. The singularity is depicted as the collapse of these spheres to a single point (the singularity). The desingularization of the singular variety is depicted as the blow up of the singular point into the bouquet of two-spheres. Cf. **Gravitational instanton.** Cf. Chap. 7.

Superstring theory. There are five versions of superstring theory:

I, IIA, IIB, Het.E, Het.O

I has both open and closed strings. **IIA** and **IIB** have only open strings. **Het.E** has $\mathcal{E}_8 \times \mathcal{E}_8$ symmetry, and only closed strings. **Het.O** has $\mathcal{SO}(32)$ symmetry, and also only closed strings. Cf. Chap. 18 and **Heterotic string theory.**

Supersymmetry. A transformation which changes **Bosons** (of integral spin) into **Fermions** (of $\frac{1}{2}$ integral spin) — and vice versa.

Symplectic space. A $2n$-dimensional real vector space on which a two-form (called a *symplectic form*) ω can be defined:

$$\omega = \sum_{A=\{1,\dots,n\}} dq^A \wedge dp_A,$$

where $\wedge$ is the wedge product. In physics, a symplectic space of dimension $6N$ is called the phase space for N particles in 3D space. In this application, q is called position and p is called momentum. Such a phase space is the geometrical structure used in Hamiltonian mechanics.

Unfolding. A family of functions F that contains a particular function f is called an unfolding of f. If F contains all the functions close to f,the unfolding is called a *universal unfolding.* The unfolding is also called a **Deformation**.

Unified field theory. A theory combing two or more **force fields** into a single force field. The theory must also give an adequate description of the **matter fields** which feel the force fields entailed in the unified theory. Matter fields interact with each other by exchanging force fields.

Grand Unified Theory (GUT) unifies the Standard Model gauge groups, $\mathcal{U}(1)\times\mathcal{SU}(2)\times\mathcal{SU}(3)$, as subgroups in a larger Lie group such as $\mathcal{SU}(5)$, or $\mathcal{SO}(10)$, or even bigger groups such as E_6, or $E_8 \times E_8$.

Vector. A set of ordered numbers called components. The number of these components is the dimension of the vector. A real vector has real numbers as components; a complex vector has complex numbers as components. Since a complex number is an ordered pair of real numbers, an n-dimensional complex vector can be regarded as a $2n$-dimensional real vector. Geometrically, a vector is an entity having magnitude and direction. We can recover the algebraic definition, if we imagine the vector tied to the origin of a coordinate system. Then the components of the vector are the coordinates of the tip of the vector.

Vector space. A space whose elements, called vectors v, form a commutative group, which by convention is considered additive so that the 0-vector is the identity element. It must also be possible to multiply v by a scalar s (i.e. a number); a law of *distribution* holds for this scalar multiplication:

$$sv_1 + sv_2 = s(v_1 + v_2).$$

Thus a vector space is defined over the set of scalars. If this set is the *real numbers*, we call the space a real vector space. If this set is the *complex numbers*, we call the space a complex vector space. Thus it is always possible to consider an n-dimensional complex vector space as a $2n$-dimensional real vector space.

Wave function. A complex function which provides a description of a quantum system. It is a function of the **observables** of the system in such a way that the square of the amplitude of the wave function is the probability for seeing a particular value of an observable. Thus the wave function is a wave of probability amplitudes. Like classical waves, wave functions are additive. Thus two wave functions (for the same system) can be added to produce another wave function. Whenever a quantum system is observed a particular aspect of the wave function, called an eigenfunction, is actualized. An alternative view of the wave function is a vector, called the **state vector.** The additivity of vectors corresponds to the additivity of wave functions. Moreover, eigenvectors correspond to the eigenfunctions. Cf. **Schrödinger equation.**

Weyl groups. Finite reflection groups which are equivalent to Coxeter's crystallographic reflection groups. However, there are Coxeter graphs corresponding to reflection groups which are not crystallographic.

Note that in the context of Lie algebras, the crystallographic reflection groups are conventionally called the Weyl group of the Lie algebra.

Cf. the historical note in Chap. 4 and also Chap. 6.

Index

SERIES ON KNOTS AND EVERYTHING

Editor-in-charge: Louis H. Kauffman *(Univ. of Illinois, Chicago)*

The Series on Knots and Everything: is a book series polarized around the theory of knots. Volume 1 in the series is Louis H Kauffman's Knots and Physics.

One purpose of this series is to continue the exploration of many of the themes indicated in Volume 1. These themes reach out beyond knot theory into physics, mathematics, logic, linguistics, philosophy, biology and practical experience. All of these outreaches have relations with knot theory when knot theory is regarded as a pivot or meeting place for apparently separate ideas. Knots act as such a pivotal place. We do not fully understand why this is so. The series represents stages in the exploration of this nexus.

Details of the titles in this series to date give a picture of the enterprise.

Published:*

Vol. 1: Knots and Physics (3rd Edition)
by L. H. Kauffman

Vol. 2: How Surfaces Intersect in Space — An Introduction to Topology (2nd Edition)
by J. S. Carter

Vol. 3: Quantum Topology
edited by L. H. Kauffman & R. A. Baadhio

Vol. 4: Gauge Fields, Knots and Gravity
by J. Baez & J. P. Muniain

Vol. 5: Gems, Computers and Attractors for 3-Manifolds
by S. Lins

Vol. 6: Knots and Applications
edited by L. H. Kauffman

Vol. 7: Random Knotting and Linking
edited by K. C. Millett & D. W. Sumners

Vol. 8: Symmetric Bends: How to Join Two Lengths of Cord
by R. E. Miles

Vol. 9: Combinatorial Physics
by T. Bastin & C. W. Kilmister

Vol. 10: Nonstandard Logics and Nonstandard Metrics in Physics
by W. M. Honig

Vol. 11: History and Science of Knots
edited by J. C. Turner & P. van de Griend

*The complete list of the published volumes in the series can also be found at
http://www.worldscientific.com/series/skae

Vol. 12: Relativistic Reality: A Modern View
edited by J. D. Edmonds, Jr.

Vol. 13: Entropic Spacetime Theory
by J. Armel

Vol. 14: Diamond — A Paradox Logic
by N. S. Hellerstein

Vol. 15: Lectures at KNOTS '96
by S. Suzuki

Vol. 16: Delta — A Paradox Logic
by N. S. Hellerstein

Vol. 17: Hypercomplex Iterations — Distance Estimation and Higher Dimensional Fractals
by Y. Dang, L. H. Kauffman & D. Sandin

Vol. 18: The Self-Evolving Cosmos: A Phenomenological Approach to Nature's Unity-in-Diversity
by S. M. Rosen

Vol. 19: Ideal Knots
by A. Stasiak, V. Katritch & L. H. Kauffman

Vol. 20: The Mystery of Knots — Computer Programming for Knot Tabulation
by C. N. Aneziris

Vol. 21: LINKNOT: Knot Theory by Computer
by S. Jablan & R. Sazdanovic

Vol. 22: The Mathematics of Harmony — From Euclid to Contemporary Mathematics and Computer Science
by A. Stakhov (assisted by S. Olsen)

Vol. 23: Diamond: A Paradox Logic (2nd Edition)
by N. S. Hellerstein

Vol. 24: Knots in HELLAS '98 — Proceedings of the International Conference on KnotTheory and Its Ramifications
edited by C. McA Gordon, V. F. R. Jones, L. Kauffman, S. Lambropoulou & J. H. Przytycki

Vol. 25: Connections — The Geometric Bridge between Art and Science (2nd Edition)
by J. Kappraff

Vol. 26: Functorial Knot Theory — Categories of Tangles, Coherence, Categorical Deformations, and Topological Invariants
by David N. Yetter

Vol. 27: Bit-String Physics: A Finite and Discrete Approach to Natural Philosophy
by H. Pierre Noyes; edited by J. C. van den Berg

Vol. 28: Beyond Measure: A Guided Tour Through Nature, Myth, and Number
by J. Kappraff

Vol. 29: Quantum Invariants — A Study of Knots, 3-Manifolds, and Their Sets
by T. Ohtsuki

Vol. 30: Symmetry, Ornament and Modularity
by S. V. Jablan

Vol. 31: Mindsteps to the Cosmos
by G. S. Hawkins

Vol. 32: Algebraic Invariants of Links
by J. A. Hillman

Vol. 33: Energy of Knots and Conformal Geometry
by J. O'Hara

Vol. 34: Woods Hole Mathematics — Perspectives in Mathematics and Physics
edited by N. Tongring & R. C. Penner

Vol. 35: BIOS — A Study of Creation
by H. Sabelli

Vol. 36: Physical and Numerical Models in Knot Theory
edited by J. A. Calvo et al.

Vol. 37: Geometry, Language, and Strategy
by G. H. Thomas

Vol. 38: Current Developments in Mathematical Biology
edited by K. Mahdavi, R. Culshaw & J. Boucher

Vol. 39: Topological Library
Part 1: Cobordisms and Their Applications
edited by S. P. Novikov & I. A. Taimanov

Vol. 40: Intelligence of Low Dimensional Topology 2006
edited by J. Scott Carter et al.

Vol. 41: Zero to Infinity: The Fountations of Physics
by P. Rowlands

Vol. 42: The Origin of Discrete Particles
by T. Bastin & C. Kilmister

Vol. 43: The Holographic Anthropic Multiverse
by R. L. Amoroso & E. A. Ranscher

Vol. 44: Topological Library
Part 2: Characteristic Classes and Smooth Structures on Manifolds
edited by S. P. Novikov & I. A. Taimanov

Vol. 45: Orbiting the Moons of Pluto
Complex Solutions to the Einstein, Maxwell, Schrödinger and Dirac Equations
by E. A. Rauscher & R. L. Amoroso

Vol. 46: Introductory Lectures on Knot Theory
edited by L. H. Kauffman, S. Lambropoulou, S. Jablan & J. H. Przytycki

Vol. 47: Introduction to the Anisotropic Geometrodynamics
by S. Siparov

Vol. 48: An Excursion in Diagrammatic Algebra: Turning a Sphere from Red to Blue
by J. S. Carter

Vol. 49: Hopf Algebras
by D. E. Radford

Vol. 50: Topological Library
Part 3: Spectral Sequences in Topology
edited by S. P. Novikov & I. A. Taimanov

Vol. 51: Virtual Knots: The State of the Art
by V. O. Manturov & D. P. Ilyutko

Vol. 52: Algebraic Invariants of Links (2nd Edition)
by J. Hillman

Vol. 53: Knots and Physics (4th Edition)
by L. H. Kauffman

Vol. 54: Scientific Essays in Honor of H Pierre Noyes on the Occasion of His 90th Birthday
edited by J. C. Amson & L. H. Kauffman

Vol. 55: Knots, Braids and Möbius Strips
by J. Avrin

Vol. 56: New Ideas in Low Dimensional Topology
edited by L. H. Kauffman & V. O. Manturov

Vol. 57: ADEX Theory: How the ADE Coxeter Graphs Unify Mathematics and Physics
by S.-P. Sirag

Printed in the United States
By Bookmasters